ICSDEMS 2019

Seyed Sattar Emamian · Timothy O. Adekunle ·
Utaberta Nangkula · Mokhtar Awang
Editors

ICSDEMS 2019

Proceedings of the International Conference on Sustainable Design, Engineering, Management and Sciences

Springer

Editors
Seyed Sattar Emamian
Mechanical Engineering Department
Universiti Malaya
Kuala Lumpur, Malaysia

Utaberta Nangkula
Linton University College
Mantin, Malaysia

Timothy O. Adekunle
University of Hartford
Hartford, CT, USA

Mokhtar Awang
Department of Mechanical Engineering
Universiti Teknologi Petronas
Seri Iskander, Perak, Malaysia

ISBN 978-981-15-3764-6 ISBN 978-981-15-3765-3 (eBook)
https://doi.org/10.1007/978-981-15-3765-3

This Springer imprint is published by the registered company Springer Nature Singapore Pte Ltd.
The registered company address is: 152 Beach Road, #21-01/04 Gateway East, Singapore 189721, Singapore

Acknowledgements

The editors would like to thank all the members of the local organizing committee who helped organize the International Conference on Sustainable Design, Engineering, Management and Sciences (ICSDEMS 2019), held in Kuala Lumpur, Malaysia. We would like to thank the colleagues and staff members at the institutions and organizations that served as partners for the international conference. Their support in organizing a successful conference has helped the editors to gather ideas and papers presented in this book. The editors are grateful to all the speakers who attended the conference and shared from their wealth of experience some exciting findings who have further propelled us to publish this book.

The editors also appreciate various people, including the production team at Springer, who helped and contributed to the creation of this book. We thank all the authors and contributors who presented at the conference and sent us their papers for peer review. The editors would like to thank and appreciate the peer reviewers for their suggestions, comments, efforts, and time spent to go over all the papers. Their contributions have helped the editors to reach final decisions on many chapters included in this book. The editors also appreciate the following organizations, including Linton University College, Universitas Indonesia, WARIS Research Group, Universiti Putra Malaysia, Collaboration with Relife Green Development, Rumah Intaran, FUSI Foundation, BUPiBE Research, and University of Hartford (USA), for their contributions and support to the overall success of the conference.

The editors appreciate the support of the leadership team of their respective institutions for the support, encouragement, and enabling environment created to prepare this book. The conference has inspired and brought the editors together from different disciplines, institutions across different countries, and continents of the world to work on this book. The creation of this book has helped us to become a formidable team. The process has been enjoyable, challenging, inspiring, and more peaceful than we ever thought. Lastly, the editors would like to thank all the team members at Springer for their unstinting support of this book, especially the production manager, the commissioning editor, and the managerial team. We are grateful for your support. We thank you all!

Timothy O. Adekunle
(on behalf of the editors)
Seyed Sattar Emamian
Utaberta Nangkula
Mokhtar Awang
(Eds.)

Introduction

This book captures the research outcomes presented during the International Conference on Sustainable Design, Engineering, Management and Sciences (ICSDEMS 2019). The international conference was held from October 16 to 17, 2019, with a focus on "Sustainable Environment." The overarching theme of the conference centered on "Towards a Sustainable Environment: Exploring the Solutions for the Future." The conference was created to provide a platform to researchers, engineers, building scientists, designers, planners, project managers, and professionals from different sectors across the globe to present, share, and exchange their research outcomes and developmental activities on all aspects of green built environment. This book showcases the efforts of the presenters and contributors regarding their research findings and how the results could promote a global partner for research collaboration in future. The conference provides an opportunity for delegates to have face-to-face discussion and thus establishes the expected research networking for participants in various specializations related to sustainable architecture, building, construction, engineering, planning, information technology, and project management. The editors carefully considered the peer reviewers' reports and comments which were sent to the presenters and contributors. The comments, suggestions, and recommendations have shaped and strengthened the quality of the chapters presented in this book.

This book comprises 29 chapters from various presenters and contributors. The first five chapters (Chapters Eco-Friendly Masonry Products for Affordable Housing—Perspective of Positive Social Impact–Service Quality Gap Analysis of Water Supply in Urban Areas) capture a range of topics from application of eco-friendly masonry products for affordable housing to conformity of the use of space and land use in a metropolis. The first section also captures the factors influencing the perception of people in the surrounding community of a village to analyze the pedestrian movements due to the operational activities in a transit station. Additionally, the first section considers a study on service quality gap analysis of water supply in a water supply company of urban areas. The first five chapters show a strong connection between the applications of eco-friendly products for affordable housing to use of land and space. The first five chapters also discuss the factors affecting people's perception of the immediate environment to understanding how operational activities affect pedestrian patterns in a transit station to understanding the service quality gap of water supply in many urban areas.

The second five chapters (Chapters Review on Research Methods in Performance-Based Building Design of High-Rise Residential Property–Analysis of Gen Y Lifestyle Based on Life Cycle on Housing Preferences in Depok City) of this book focus on topics such as review of research methods in performance-based multi-story residential buildings, an investigation on non-destructive tests of old concrete specimens under different load and unload conditions. Other topics include the use of polyethylene terephthalate plastic bottle waste for modification of asphalt concrete mix, analysis of land carrying capacity and population capacity around a toll road of a metropolitan area, and analysis of a lifestyle based on life cycle on housing preferences in a city. The topics would help readers understand how the use of materials and waste products can help attain a sustainable environment due to a paradigm shift in our lifestyle.

The third five chapters (Chapters Architect's Knowledge and Perception to Apply Green Building Aspect in Design–Pesanggrahan River Management in the Recent Times, The Anthropocene Era: A Case Study of Sangga Buana Urban Forest, Jakarta) focus on collection of work on topics including architect's knowledge and perception to apply green building concept in design, analysis of the impact of urban development on river water quality: a case study of a metropolitan area, spatial assessment of micro-hydro power plant in subsidized housing in a community, cost–benefit analysis of hybrid retaining walls construction, and river management in the recent times: a case of an urban forest. The third five chapters emphasize the importance of different aspects of sustainable environment, including river water quality, spatial assessment of hydro power plant, and cost–benefit analysis of green built environment. The chapters would also help readers understand some overarching issues regarding river management in urban forests.

The fourth five chapters (Chapters Simulation of Damage due to Alkali-Silica Reaction in a Concrete Model at the Macroscale Level–Sustainability in Architectural Conservation of Heritage Building: A Qualitative Approach) discuss a range of topics from simulation of damage due to alkali–silica reaction in a concrete model at the macroscale level to combination of LEM and FEM analysis for stability of concrete cantilever retaining wall. The fourth five chapters also contain other topics such as application of waste low-density polyethylene (LDPE) as bitumen modifier in asphalt concrete-binder course (AC-BC) mix, comparative study of energy efficiency and thermal comfort technology: case studies of eco-friendly studio and house, and a qualitative approach to sustainability in architectural conservation of heritage buildings. This part of the book would help readers capture how technology has helped promote sustainable environment from the perspective of engineering, building science, and architectural conservation.

The fifth five chapters (Chapters Analysis of Road Traffic Services due to the Operation of Cibubur Greater Jakarta Light Rail Transit Station–Bearing Strength of Concrete Blocks with External Wrapping of Carbon Fibre Reinforced Plastic (CFRP)) of the book capture and present some interesting topics such as analysis of road traffic services due to the operational activities at a light rail transit station and analysis of pedestrian scenarios in public spaces: a case study of a metro light rail transit station. Other topics include changes of facades in a building with commercial function in an urban area, bearing strength of reinforced concrete blocks axially loaded through different sizes of steel plates, and bearing strength of concrete blocks with external wrapping of carbon fiber-reinforced plastic (CFRP). The topics present in Chapters "Analysis of Road Traffic Services due to the Operation of Cibubur Greater Jakarta Light Rail Transit Station"–"Bearing Strength of Concrete Blocks with External Wrapping of Carbon Fibre Reinforced Plastic (CFRP)" would help readers capture the various developmental activities regarding the aspects of sustainable built environment in light rail transit stations, building facades, reinforced concrete structures, and carbon fiber-reinforced plastic.

The last four chapters (Chapters "Mechanical properties of Engineered Cementitious Composite (ECC): An Overview"–"Impact of Urban Configurations on Airflow: Tropical Context Study") discuss a selection of topics that range from an overview of mechanical properties of engineered cementitious composite (ECC) to a comparative study of solar heat transmission through single-skin façade (SSF) and naturally ventilated double-skin façade under a hot and humid climate. The last four chapters also capture topics on responsible waste management of micro, small, and medium enterprises (MSMEs) toward sustainable development in a case study city and impact of urban configurations on airflow: tropical context study. Chapters "Mechanical properties of Engineered Cementitious Composite (ECC): An Overview"–"Impact of Urban Configurations on Airflow: Tropical Context Study" provide insight on various ways we can promote a sustainable environment through use of single-skin façade and naturally ventilated double-skin façade. This part of the book also includes information on responsible waste management at various levels toward attaining sustainable development in different locations and contexts.

The editors hope the collections of chapters in this book will help readers, including researchers, academia, and industrial professionals, reach out to the presenters and contributors for continued networking and collaboration toward attaining a sustainable environment. We are optimistic that readers would find the research findings in this book helpful and promote future discussion on how some of the research projects could be further developed. It is our hope the collection of work in this book would help provide answers or hints to some unanswered questions in the fields of sustainable architecture, building technology, construction, engineering, planning, information technology, and project management. We hope you enjoy reading the book. We also hope we can build on this work in the future.

Timothy O. Adekunle
(on behalf of the editors)
Seyed Sattar Emamian
Utaberta Nangkula
Mokhtar Awang
(Eds.)

Contents

Eco-Friendly Masonry Products for Affordable Housing—Perspective of Positive Social Impact 1
R. Ramesh Nayaka, U. Johnson Alengaram, Mohd Zamin Jumaat, and K. Balakrishna Rao

Conformity of the Use of Space and Land Use in North Jakarta Administrative City 13
Galih Jati Utomo, Hafid Setiadi, and Rudy P. Tambunan

Factors Affecting the Perception of the Surrounding Community in Maintaining the Sustainability of the Setu Babakan Betawi Cultural Village in South Jakarta 27
Dwinanto Suberlian and Chotib

Analysis of Pedestrian Movements Due to the Operations of Cibubur LRT Station 33
Wiby Adhitya Prayoga and Jachrizal Sumabrata

Service Quality Gap Analysis of Water Supply in Urban Areas: Water Supply Company, Bogor Regency 47
Armeinita Octavia and Chotib

Review on Research Methods in Performance-Based Building Design of High-Rise Residential Property 59
Sulfiah Dwi Astarini and Christiono Utomo

A Study of Non-destructive Test (NDT) Results on Old Concrete Specimens Under Loaded and Unloaded Conditions and Their Crushing Strength 67
Ajmal Paktiawal and Mehtab Alam

The Utilization of Polyethylene Terephthalate Plastic Bottle Waste for Asphalt Concrete Mix Modification 83
Amalia Firdaus Mawardi, Machsus Machsus, Rachmad Basuki, Mohamad Khoiri, Sukobar, and Ahmad Faqihul Muqoddam

Analysis of Land-Carrying Capacity and Population Capacity Around Manado–Bitung Toll Road, North Sulawesi Province 95
Chairitha Rahmasani, Hafid Setiadi, and Komara Djaja

Analysis of Gen Y Lifestyle Based on Life Cycle on Housing Preferences in Depok City 109
M. Akmal Farraz, Lita Sari Barus, Khoirunurrofik, and Joko Adianto

Architect's Knowledge and Perception to Apply Green Building Aspect in Design 119
Togu Riotama, Andrean Eka Lucianto, and Herdis Herdiansyah

Analysis of the Impact of Urban Development on River Water Quality Case Study of the Pesanggrahan River 127
Muslim Aminuddin and Chotib

Spatial Assessment of Micro-hydropower Plant in Subsidized Housing Case: Perumnas Parung Panjang 141
Andrean Eka Lucianto and Herdis Herdiansyah

Cost-Benefit Analysis of Combined Retaining Walls Construction 149
Mohd. Sukry Mohamed and Samira Albati Kamaruddin

Pesanggrahan River Management in the Recent Times, the Anthropocene Era: A Case Study of Sangga Buana Urban Forest, Jakarta 155
Neli Triana, Linda Darmajanti, Yophie Septiady, Khoirunurrofik, and Komara Djaja

Simulation of Damage Due to Alkali–Silica Reaction in a Concrete Model at the Macroscale Level 169
Zarina Itam, Agusril Syamsir, Salmia Beddu, and Nur Liyana Mohd. Kamal

Combination of LEM and FEM Analysis for Stability of Concrete Cantilever Retaining Wall 179
Rohaya Alias, Siti Jahara Matlan, and Nur Asmaliza Mohd. Noor

Use of Waste Low-Density Polyethylene (LDPE) as Bitumen Modifier in Asphalt Concrete-Binder Course (AC-BC) Mix 187
Sukobar, Machsus Machsus, Rachmad Basuki, Amalia Firdaus Mawardi, and Moh. Firli Firdausi

Comparison of Energy Efficiency and Thermal Comfort Technology: Case Studies of Tsinghua Eco Studio & S11 House 195
Tan Jia Moon, Nangkula Utaberta, Chia Lin Lin, and Nayeem Asif

Sustainability in Architectural Conservation of Heritage Building: A Qualitative Approach 201
Yeo Shu Han, Nangkula Utaberta, Nayeem Asif, and Norizan Daud

Analysis of Road Traffic Services Due to the Operation of Cibubur Greater Jakarta Light Rail Transit Station 211
Rakhmat Shafly Syabana and R. Jachrizal Sumabrata

The Analysis of Pedestrian Scenarios in Public Spaces (Case: Harjamukti LRT Station, Cibubur) 221
Lucky Steffano Samuel Situmorang and R. Jachrizal Sumabrata

The Changes of Facades in Jengki Building with Commercial Function in Yogyakarta 229
Noor Zakiy Mubarrok and Sidhi Pramudito

Bearing Strength of Reinforced Concrete Blocks Axially Loaded Through Various Sizes of Steel Plates 241
Mohd Raizamzamani Md Zain, Norrul Azmi Yahya, Siong Wee Lee, Chai Lian Oh, and Anizahyati Alisibramulisi

Bearing Strength of Concrete Blocks with External Wrapping of Carbon Fibre Reinforced Plastic (CFRP) 249
Norrul Azmi Yahya, Norliyati Mohd Amin, Mohd Raizamzamani Md Zain, Oh Chai Lian, and Lee Siong Wee

Mechanical Properties of Engineered Cementitious Composite (ECC): An Overview 259
Nurmazidah Abdullah Zawawi, Chai Lian Oh, Siong Wee Lee, Mohd Raizamzamani Md Zain, and Norrul Azmi Yahya

Comparative Study of Solar Heat Transmission Through Single Skin Façade (SSF) and Naturally Ventilated Double Skin Façade Under Malaysian Climate 265
Olutobi Gbenga Ayegbusi, Abdullah Sani Ahmad, and Yaik Wah Lim

Responsible Waste Management of Micro, Small, and Medium Enterprises (MSMEs) Toward Jakarta Sustainable Development 273
David Ronald Tairas, Rosnaini Daga, Nur Fatwa, Lin Yola, and S. Basir

Impact of Urban Configurations on Airflow: Tropical Context Study 281
Lin Yola, Siong Ho Chin, and Komara Komara Djaja

About the Editors

Dr. Seyed Sattar Emamian received his Doctorate degree in Mechanical Engineering from Universiti Teknologi PETRONAS (UTP), Malaysia in 2018. He is a member of Board of Engineers Malaysia (BEM). He has published several ISI and Scopus indexed journals/proceedings as well as book chapters. His research area are welding engineering, manufacturing systems engineering and materials science. He is currently a research fellow at Universiti Malaya, Malaysia.

Professor Timothy O. Adekunle currently works as a full-time faculty in the Department of Architecture at the University of Hartford, USA. He holds a Ph.D. in Architecture from the University of Kent, UK. Professor Adekunle is an alumnus of the University of Nottingham, UK, where he obtained his postgraduate degree (M.Arch.) in Architecture. Professor Adekunle previously worked as an Assistant Lecturer at the School of Architecture, University of Kent, UK, before he relocated to the United States. He is an international scholar who has studied and worked in three different continents. He has won numerous local and international awards. Professor Adekunle is an expert in the field of sustainable architecture, innovative building technology, and integrated design, completing many research projects on thermal environments of different buildings, including a recent project on an innovative sustainable school building development in the Northeast region of the United States. He has published his research outputs as peer-reviewed articles in numerous top-ranking journals, has presented peer-reviewed papers in professional conferences, and is a reviewer for top journals (Energy and Buildings, Automation in Construction, Architectural Science Review, Building and Environment, Sustainable Cities and Society, Energies, Building Research and Information among others). He is an implementation partner for a National Science Foundation (NSF) nation-wide funded research project on passion-driven data. Professor Adekunle is a member of many professional and academic organizations, including a Fellow of the UK Higher Education Academy (now Advance HE), an Associate Member of the American Institute of Architects (AIA), a member of the Society of Building Science Educators (SBSE), and a full member of the Chartered Institute of Building (CIOB).

Utaberta Nangkula is an International Writer, Professor and Researcher who has written 22 academic research books, 52 chapters in books, more than 400 papers in indexed journals (SCOPUS & WoS), 139 papers in indexed proceedings and over 30 articles in newspaper of Architecture in Indonesia and Malaysia. Born in Jakarta, Nangkula obtained his Doctor of Philosophy and Master of Architecture degree from Universiti Teknologi Malaysia in Johor Bahru as the Best Graduate of Architecture in 2005 and 2009.

Currently he is a Research Fellow, at Halal Product Research Institute (IPPH), Universiti Putra Malaysia (UPM), where he is also Professor of Architecture and Director of Research Development Centre (RDC), Linton University College. Before joining this university, he served as Associate Professor at Faculty of Design and Architecture, UPM (2012–2018). Senior Lecturer (Assistant Professor) at National University of Malaysia-UKM (2009–2012), Deputy Dean and founder of School of Architecture and Design at Legenda Education Group (2006–2009) and lecturer at SPACE program, Universiti Teknologi Malaysia and University of Indonesia. Nangkula was once visiting Lecturer at the State Islamic University of Malang (UIN Malang-2009) and University of Muhammadiyah

Surakarta (UMS-2005). He is also an Associate Research Fellow at various Centers of Excellence in Malaysia and Indonesia including the Honorary Advisor at Relife Green Development, Expert Committee of the Dewan Bahasa dan Pustaka Malaysia, Senior Advisor at Center for Islamic Architecture (Islamic State University of Malang), Center for Indonesian Reform and Center for Islamic Architecture Studies (Muhamadiyah University of Surakarta).

In Book Writing, Nangkula is the recipient of the National Book of Malaysia Award for Academic, General and Creative Books for three consecutive years in 2014, 2015 and 2017 awarded by Prime Minister of Malaysia and Sultan (king) of Selangor. His academic and popular books have been translated into Indonesian and English language and become one of the major reference books in the teaching and research of Islamic Architecture in Indonesia and Malaysia. He is a passionate academician and is often asked to give keynote lecture on Mosque design and Islamic Architecture in many conferences, seminars and symposia in Indonesia, Malaysia and various countries in Asia and Europe.

As an academic since 2009, he has graduated 11 Ph.D. and 17 Master of Architecture (by research) students. Lead in 20 major research projects in Malaysia including the Fundamental Research Grant, Trans Disciplinary Research Grant (TRGS), Flood Disaster Management, Group Putra Initiative Putra, Impact Putra, Putra Grant, Graduate Putra, Graduate Research University (GUP), Community-University Grant and Industrial University Grant worth RM1,195,763.00 and involved in many Malaysian Government Major Research Projects such as Permata Project, Lake Chini Biosphere Research and Flood Disaster Management worth more than RM2,512,958.00. He is also involved in several interdisciplinary and industrial research projects since 2010, bringing him the High Impact Industry and Community Award from the UPM Vice-Chancellor for the Design of Indonesian School of Kuala Lumpur in 2016.

Ir. Dr. Mokhtar Awang is an Associate Professor at Mechanical Engineering Department, University Teknologi Petronas and Head of Computational Modeling Group. He has obtained his Ph.D. degree in 2007 from West Virginia University, USA. He is a Professional Engineer registered with Board of Engineers Malaysia (BEM) as well Chartered Engineer with IMarEST, UK. He has more than 6 years experience working with industry and 17 years in academia. His research focus areas are Computational modelling using commercial software ANSYS, ABAQUS and Altair Hyperworks. He has published more than 80 indexed journals and a co-author of *Finite Element Modelling of Nanotube Structures: Linear and Non-linear Models* and 5 other edited books, mostly published by Springer.

Eco-Friendly Masonry Products for Affordable Housing—Perspective of Positive Social Impact

R. Ramesh Nayaka, U. Johnson Alengaram, Mohd Zamin Jumaat, and K. Balakrishna Rao

Abstract Although many countries have the wish to assess the social impact, it is not clear how to evaluate societal quality, exclusively for basic and strategic research. Besides, it is very essential for a young researcher, engineer, and scientist to understand the importance of societal impact for research. The major contribution to infrastructure development and housing comes through the construction industry. However, due to extensive usage of natural resources as conventional materials in construction, they have led to a rise in environmental problems and a rise in cost of materials which has impact on social cost. Improper waste management causes the health impact rise due to environmental problem. Hence, it is important to translate the ideas to fill the gap between solid waste and construction industry through advancing the understanding of fundamentals and developing solutions to problems of social significance. Therefore, this study attempted to utilize industrial by-products (IP's) to develop masonry products. In this research, the palm oil industry by-products such as palm oil clinker powder (POCP) was used as an example to address the social impact. The basic properties of POCP were investigated to meet the standard limit, and engineering performance was investigated to evaluate the masonry mortar containing POCP as cement replacement. Eventually, the reduction of carbon emission and cost were evaluated to assess the positive impact on society. The investigation results revealed that adequate engineering performance achieved by POCP incorporated masonry mortar. In addition, appreciable reduction of carbon emission (32%) and cost (20%) were noticed due to the incorporation of IP's. Further, the landfilling was reduced by diverting IP's that were utilized in masonry products. The new masonry mortar or masonry products can be used in masonry construction and can be used to alleviate the inadequate supply of affordable housing.

R. R. Nayaka (✉) · K. B. Rao
Department of Civil Engineering, Manipal Institute of Technology (MIT), Manipal Academy of Higher Education (MAHE), Manipal 576104, Karnataka State, India
e-mail: ramesh.nayaka@outlook.com; ramesh.n@nitw.ac.in

R. R. Nayaka · U. J. Alengaram · M. Z. Jumaat
Centre for Innovation Construction Technology (CICT), Department of Civil Engineering, Faculty of Engineering, University of Malaya, 50603 Kuala Lumpur, Malaysia

S. S. Emamian et al. (eds.), *ICSDEMS 2019*,
https://doi.org/10.1007/978-981-15-3765-3_1

Keywords Societal impact · Affordable Housing · Palm oil clinker powder (POCP) · Masonry products · Carbon emission · Cost-efficiency

1 Introduction

Inadequate supply of affordable housing for the growing population in the Asia region has caused a rapid increase in construction activities which has a very high social impacted as the matter of basic need. Therefore, adopting alternatives to meet the challenge of bridging the affordability gap is significant. For instance, 89% of total wastes in Malaysia goes to landfilling which pollutes the soil and water causes environmental problems, and this waste generation has been forecasted to increase by 65% by 2020 [1, 2]. Although, Malaysia stands in the first place among palm oil exporters and the second in palm oil production worldwide [1]. Eventually, an enormous quantity of palm oil wastes are produced as in the form of empty fruit bunches (EFB), mesofibres, palm shells, etc., which are posing disposal problems [3]. However, 85% of waste is reused as fossil fuel to generate flux to produce palm oil [4]. Even after using as fossil fuel, the combusted waste about 25% is remained as in the form of palm oil clinker (POC) and palm oil fuel ash (POFA) which goes to landfilling and causes air and soil pollution [5]. Therefore, diversion of palm oil waste by-products is essential and could be an alternative environmental-friendly building material to utilise in masonry mortar to preserve the natural resources.

Previous research studies have reported on the utilization of palm oil mills by-products in mortar and concrete. Extensive research work on using of palm oil shell in blast resistance concrete has been highlighted [6], and previous works on using of POC as coarse aggregate in self-compacting concrete, high strength concrete and lightweight concrete were noticed [7–10]. In a study, Karim et al. [9] revealed that POCP is a pozzolanic material which is better than palm oil fuel ash (POFA). Nayaka et al. [7, 11] studied to replace cement with POCP to minimize CO_2 and cost in masonry mortar without compromising the minimum requirement of mortar. The results revealed that POCP could be used up to 40% which reflects on 32% reduction of CO_2 and about 20% reduction of cost. Hence, the challenge in diverting waste from landfilling to produce potential building materials would be remedied for the environmental problem and supporting society to afford housing. In this context, the research is investigated on the utilization of POCP as a binder to replace cement. The utilization of POCP will have an influence on cost-saving and reduce CO_2 emission [12]. In addition, the research addresses the societal impact necessary to investigate the economic and environmental vulnerability of POCP without compromising the engineering properties. Hence, this study attempted to evaluate the economic and environmental vulnerability through indices developed by Kanadasan and Razak [13], namely engineering environmental index (EEI) and engineering economic index (ECI).

2 Methods

The objectives in this research were, firstly, to utilize local industrial by-products, i.e. POCP to produce eco-friendly masonry products. Secondly, to analyse the performance of masonry products, and finally, to evaluate carbon footprint and cost of eco-friendly materials for environmental and social benefits. Figure 1 presents the method adopted in this study to attempt the challenges and get solutions through this research which contribute to powering the green cities and has a positive impact on society in terms of economy and environment.

Figure 2 presents the conventional materials being used in masonry products and available alternative materials from industrial by-products to replace the conventional materials. In addition, it shows that the masonry products could be produced using these industrial by-products. However, extensive research work is required to attempt all the industrial by-products and to produce a variety of masonry products. Therefore, this research presents only an attempt of using POCP as cement replacement in masonry mortar to show the positive societal impact. Similar could be followed for further research work on other industrials by-products to produce masonry products.

The type "S" cement lime mortar is selected in this study which is more suitable for laying bricks and blocks and could be used for rendering work with lower compressive strength and improved bonding property. The mix proportion of the cement lime mortar is 1:0.5:4.5 (Binder: Lime: Sand). The water (based on w/b ratio) was

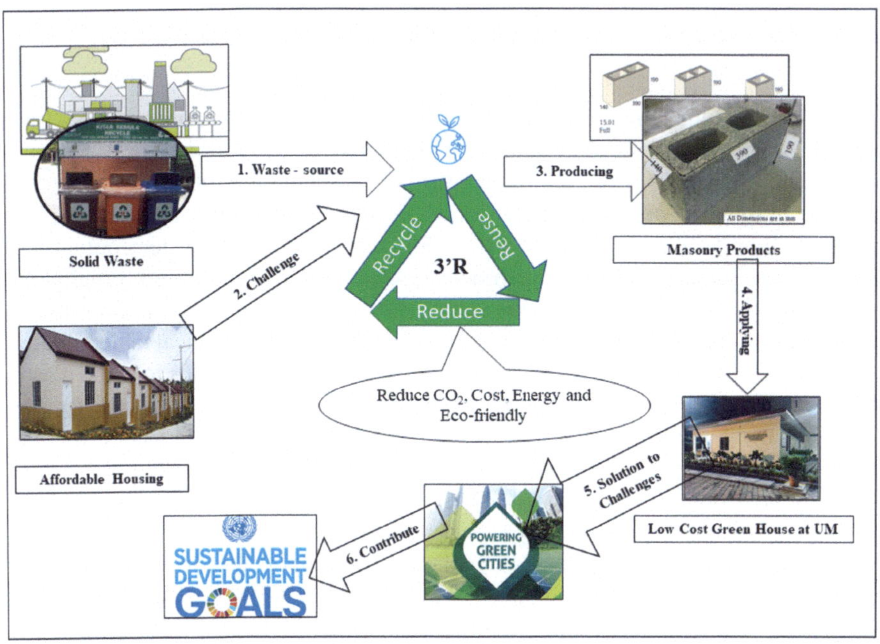

Fig. 1 Schematic diagram for attempting challenges for the solutions

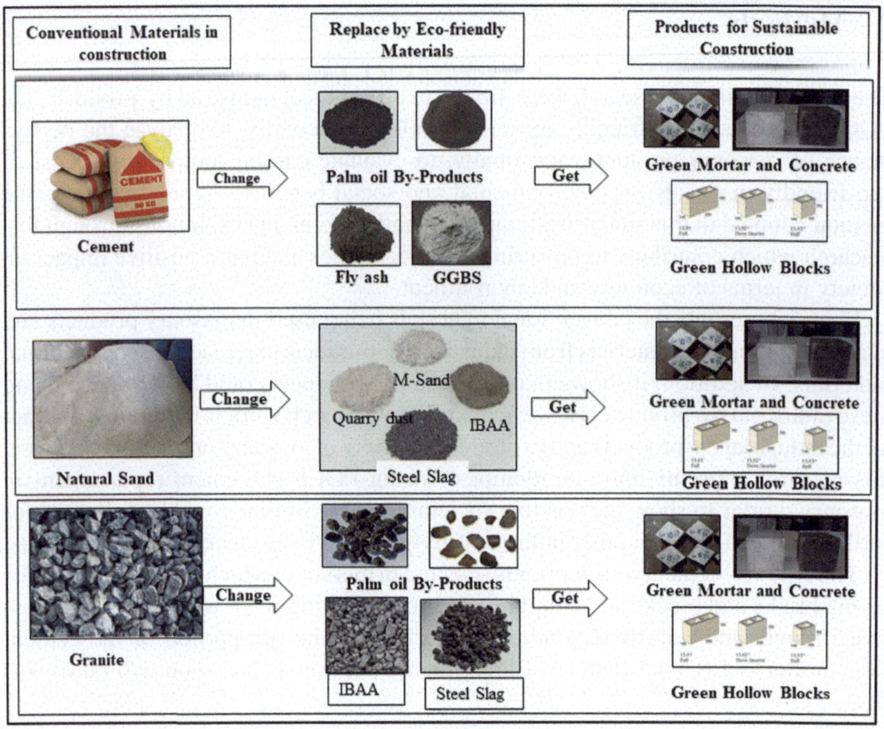

Fig. 2 Possible alternative materials to replace conventional material to produce masonry products

adjusted to meet the consistency of 110 ± 5% based on ASTM standard [18]. First of all, OPC, hydrated lime, and sand were blended in a mortar mixer for 3–5 min to obtain a homogeneous blend and then added to the blend and blended for another 2–3 min to obtain a coherent mortar paste. Secondly, the mortar paste was tested for slump flow to attain a value of 110 ± 5%, and accordingly, water content was adjusted, and subsequently, the fresh density of mortar was measured. Finally, the paste was taken from the mixing bowl and deposited into 50 mm cubic moulds in two layers in the mould and each layer was compacted by using mechanical table vibrator. Furthermore, the moulds were let for curing by covering plastic sheets to avoid plastic shrinkage of the mortar due to evaporation of moisture from mortar paste. After 24 h of initial curing, the specimens were demoulded and transported to wet curing unit for further curing in the laboratory environment with the varying temperature of 28–32 °C and relative humidity of 70–80%. Similar steps were followed for the mixes that were prepared with POCP substitution. The masonry mortar mixes were prepared by substitution of POCP to replace cement up to 80% with a gradual of increase of 10% for every mix prepared. The ideal mixture was selected based on meeting the required strength of 12.40 MPa based on ASTM C270-14 [15]. In the investigation, the hardened properties of every mix were determined by the hardened density and compressive strength based on ASTM C109-14 [18]. The reduction of

Table 1 Fresh properties of mortar with respect to POCP replacement level

Mix ID	W/C ratio	W/B ratio	A/B ratio	Slump flow (mm)	Fresh density (kg/m^3)
CL	0.5	0.46	3.83	108.00	2225
CLP10	0.59	0.49	3.83	115.00	2185
CLP20	0.67	0.57	3.83	113.25	2135
CLP30	0.67	0.57	3.83	106.50	2120
CLP40	0.67	0.57	3.83	111.75	2095
CLP50	0.67	0.57	3.83	109.50	2060
CLP60	0.70	0.60	3.83	112.50	2035
CLP70	0.70	0.60	3.83	108.50	2022
CLP80	0.70	0.60	3.83	110.75	2015

carbon emission and cost reduction were evaluated. In addition, engineering economical index (ECI) and engineering environmental index (EEI) were selected from the previous conducted study [13, 14]. Indices were determined, and a comparison was drawn between conventional and environmentally roundly masonry mortars of the ideal mixture.

3 Results and Discussion

3.1 *Fresh Properties*

The required slump flow of conventional mortar reached a w/b ratio of 0.40, while the POCP replaced the w/b ratio in place of cement as the POCP content increased due to the presence of unburnt carbon in the POCP and irregular particle sizes of powder [12, 17]. The w/b ratio rose from 0.49 to 0.60 with respect to POCP replacement with cement is presented in Table 1.

3.2 *Hardened Density*

Figure 3 shows the hardened density of mortar prepared using POCP, the cement replacement increased in the mortar when POCP was reduced the hardened density. It was observed that the hardened density of POCP mortar declined from 2375 kg/m^3 at 0% of POCP to 2025 kg/m^3 at 80% of the POCP. When high volume of POCP (80%) is used in mortar, then the hardened density was decreased by 14.74% as compared to control mortar mix. There were decreases in density as the POCP content increased due to lower specific gravity, and coarser particles of POCP influenced the developing voids which resulted in the reduction of hardened densities [7, 16].

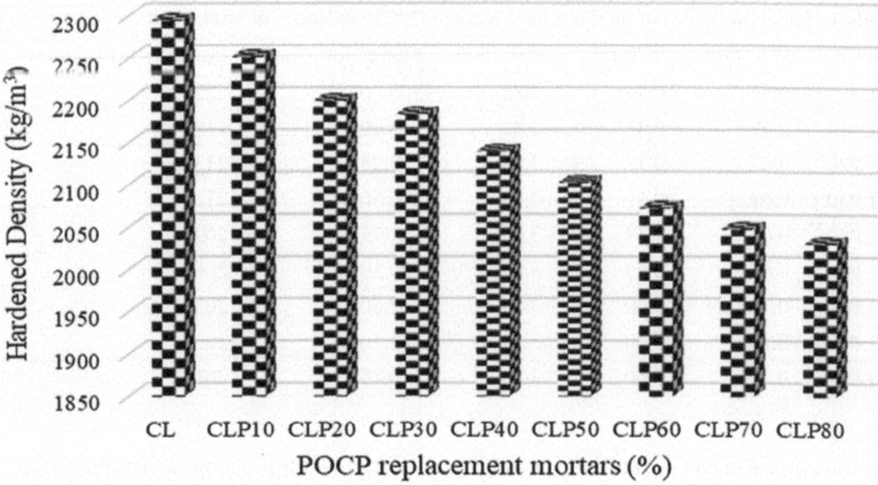

Fig. 3 Effect of POCP replacement level on hardened density

3.3 *Compressive Strength*

Figure 4 demonstrates the impact of replacing POCP on masonry mortar's compressive strength. The POCP is replaced up to 80% with cement, and the strength of the mortar mix with 80% of POCP was considerably attained less at 28 days of curing period; however, only 10% of POCP replacement was attained higher strength than

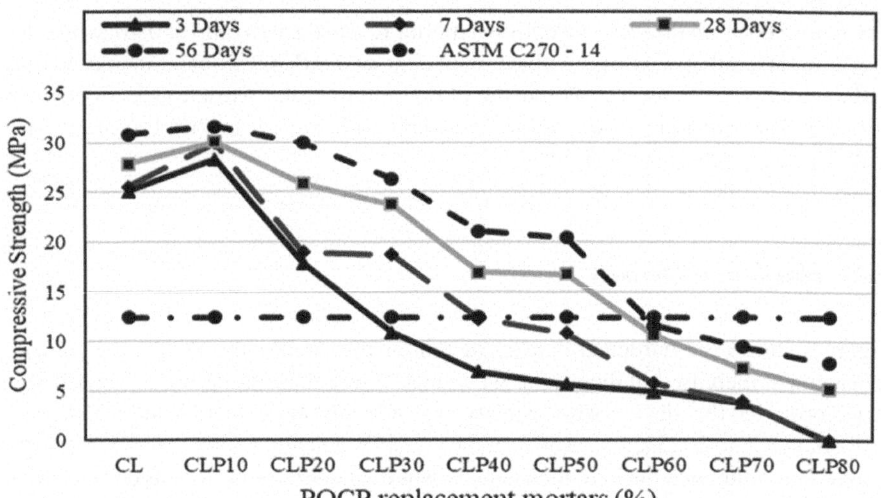

Fig. 4 Influence of POCP replacement level on compressive strength

the conventional mortar due to influence of pozzolanic effect from rich silica content. In the meantime, CLP50 achieved 58% of the strength control and more than the required compressive strength of 12.40 MPa. However, CLP60 and CLP70 obtained compressive strength of about 35 and 18% of the strength of CL which is lower than requisite strength of mortar for structural purpose; therefore, CLP60 and CLP70 could be used for the non-structural application. It was noteworthy to observe that the strength of mortar had attained the value of 28 MPa at 28 days of curing, and merely 10% of POCP replacement in mortar achieved higher compressive strength of 29.20 MPa compared to control. The enhanced strength in mortar containing 10% of POCP could be attributed by higher silica content, and coarser particles helped to develop good bonding between paste and aggregates resulted in fewer voids [17]. However, the compressive strength of masonry mortar decreased as the POCP content increased from 20% onwards due to the impact of dilution of POCP, and reaction to hydration also reduced due to less cementitious content impacted on weaker bonding between aggregate and binders; therefore, it led to reduction of compressive strength. However, in accordance with requisite strength of 12.40 MPa for masonry mortar the optimum of 40% of POCP could be recommended to use in masonry mortar as far as strength is concerned [11, 15] and considering variation factors.

3.4 *Structural Efficiency (S.E.)*

The connection between compressive strength and hardened density was determined by the structural efficiency (S.E.) of masonry mortar. In Fig. 5, it can be observed that S.E. values of the obtained control mortar were about 0.012, and in 10% of POCP replacement, it was noticed that the improved S.E. might be attributed due to rich

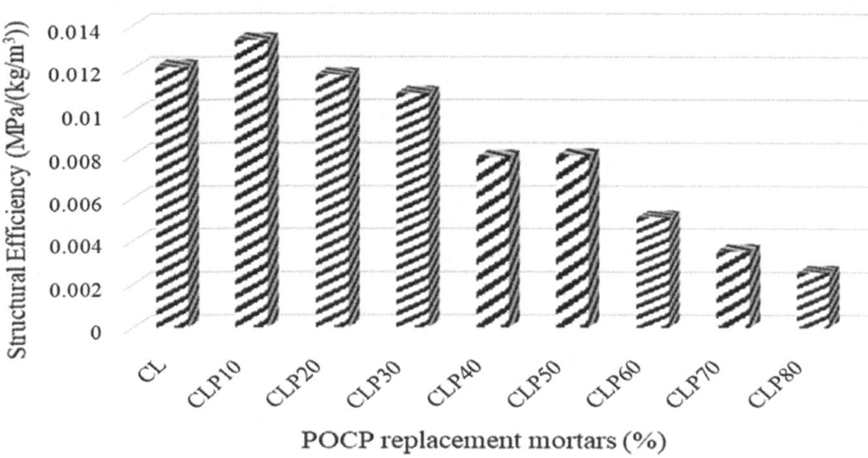

Fig. 5 POCP replacement level on the structural efficiency of mortar

silica content and presence of gypsum in POCP particles which acted as retarders to improve the strength and density; hence, that result reflected on S.E. Furthermore, the S.E. was decreased substantially as POCP content increased due to dilution effect and reduction of heat evolution rate [2, 13].

3.5 Reduction of Carbon Emission and EEI

Table 2 presents the total carbon emission and engineering environmental index (EEI) values for masonry mortar in which cement was replaced by POCP as a binder from 0 to 80%. The total carbon emission of control masonry mortar obtained 0.4085 tCO_2-e/m^3. The carbon emission of masonry mortar from 10% was about 0.3751 tCO_2-e/m^3 which implies the reduction of about 8%, and for 80% replacement, the carbon emission was reduced by 65%. However, according to requisite strength, the 40% of POCP mix was the ideal POCP mix which was based on requisite strength which reduced the carbon emission about 32%. This reduction of carbon emission could be attributed to the non-combustion process of POCP which was collected and directly grounded without any combustion. Evaluation of integrated engineering element through EEI was observed, and the obtained EEI value increased up to 40% of the POCP replacement and afterwards was decreased. The EEI value was enhanced by 23% in POCP replacement of 40% as compared control mix which indicates the POCP's capacity to generate significant engineering performance without forgoing the environmental benefits. This outcome represents the contribution to be made in increasing the use of POCP with respect to sustainability criteria [2, 7, 13].

Table 2 Total carbon production, reduction of CO_2, and EEI with respect to POCP replacement level

Mix ID	Total e-CO_2 of mortar	Reduction of CO_2 (%)	EEI
CL	0.4085	0	0.02968
CLP10	0.3751	8	0.03562
CLP20	0.3426	16	0.03424
CLP30	0.3091	24	0.03515
CLP40	0.2766	32	0.02857
CLP50	0.2431	40	0.03282
CLP60	0.2097	49	0.02431
CLP70	0.1771	57	0.019959
CLP80	0.1437	65	0.017717

Table 3 Total cost of mortar, cost reduction (%), and ECI with respect to POCP replacement level

Material (kg/m^3)	Total cost (RM/m^3)	Total cost (USD/m^3)	Cost reduction (%)	ECI
CL	312.64	73.05	0.00	3.88E−05
CLP10	296.85	69.36	5.05	4.50E−05
CLP20	281.06	65.67	10.10	4.17E−05
CLP30	265.26	61.98	15.15	4.10E−05
CLP40	249.47	58.29	20.20	3.17E−05
CLP50	233.68	54.60	25.26	3.41E−05
CLP60	217.89	50.91	30.31	2.34E−05
CLP70	202.10	47.22	35.36	1.75E−05
CLP80	186.30	43.53	40.41	1.37E−05

3.6 Cost-Saving and ECI

Table 3 presents the total cost of mortar determined for 1 m^3 of mortar, the cost reduction due to the incorporation of POCP, and engineering economical index (ECI) which shows the relationship between engineering performance and cost. In this study, POCP value was considered to be "zero cost" material. The price of materials was considered based on market values. It was found that the total cost of 1 m^3 was about RM 312.64 ($73.05) per m^3 for mortar. The total cost of CLP80 was obtained about RM 186.30 ($43.53) which was reduced about 40.41%, and for the ideal mix, the cost reduction was about 20.21%. This shows substantial building cost-savings on a wider scale. The ECI was carried out by integrating the price factor with the characteristics of the engineering. In addition to the economic advantages, the findings provide a useful indication to assess POCP's beneficial contribution to the hardened system, besides the economic advantages. For 10% of POCP mix, which is greater than control mortar, the ECI was acquired around 4.50×10^{-5}. This excellent ECI value demonstrates a healthy connection between the engineering characteristics and cost. As shown in Table 3, by integrating 40% of POCP, nearly 20.20% of the price can be decreased or saved. The cement industry is one of the biggest contributions of carbon emission to environment worldwide. The excellent correlation shows the construction industry balanced between cost and environmental elements [2, 7, 13].

4 Conclusions

Utilizing a high volume of POCP in masonry mortar for showcasing positive societal impact has been thoroughly researched in this study and concluded as follows:

- The POCP replacement up to 40% with cement is recommended to be used for masonry mortar with reference to achievement of requisite strength of 12.40 MPa.

- The replacement of cement by POCP at 40% achieved mean S.E. about 0.008 MPa/kg/m^3 and produced about 75% of control mortar. This indicates that POCP mortar is lighter than control mortar and there is a good bonding between strength and density.
- The total carbon emission was reduced by 32% for 40% replacement of POCP and could be reduced by 65% for 80% of POCP substitution for cement in masonry mortar. Reduction of carbon emission and cost have proven that POCP mortar is environmentally friendly and more cost-effective than the control mortar.
- The EEI value for CLP40 improved about 23% compared to CL, which indicates the ability of OCP to achieve engineering performance without aforementioned environmental benefits. The ECI was obtained about 4.50×10^{-5} for 10% of CLP10 which is higher than control mortar. This good value of ECI shows a good relationship between engineering properties and cost. As observed in Table 3, the cost was reduced by about 20.20% by using 40% of POCP.

In other words, powder forms of IP's such as POCP, fly ash, POFA powder, rice husk ash, and bagasse ash are reactive and these could be used to replace cement to produce eco-friendly products due to carbon-free characteristic as there is no thermal treatment involved in processing industrial by-products into masonry products; therefore, no carbon emission is produced that would help to save the environment and less energy is required to process the materials that would save the energy. Industrial by-products which are used to replace the aggregates such as palm oil clinker, oil palm shell, steel slag, and incinerated bottom ash aggregates are non-reactive. Besides, the cost of these industrial by-products is very low such as their transportation cost; therefore, they are suitable for low-cost housing that could be affordable to low-income group community. Overall, it could be concluded that adaptation of industrial by-products in the construction industry has a more significant societal impact on the environment, cost reduction, and effective solid waste management.

Acknowledgements The financial support from Manipal Academy of Higher Education (MAHE) Manipal, India is acknowledged. The research fund under the UMCIC Prototype Grant RU007C-2017H through UMCIC, University of Malaya to conduct this research work is acknowledged.

References

1. Yusoff, S. (2006). Renewable energy from palm oil—Innovation on effective utilization of waste. *Journal of Cleaner Production, 14,* 87–93.
2. Nayaka, R. R., Alengaram, U. J., Jumaat, M. Z., Yusoff, S. B., & Ganasan, R. (2019). Performance evaluation of masonry grout containing high volume of palm oil industry by-products. *Journal of Cleaner Production, 220,* 1202–1214.
3. Islam, M. M. U., Mo, K. H., Alengaram, U. J., & Jumaat, M. Z. (2016). Mechanical and fresh properties of sustainable oil palm shell lightweight concrete incorporating palm oil fuel ash. *Journal of Cleaner Production, 115,* 307–314.
4. Awalludin, M. F., Sulaiman, O., Hashim, R., & Nadhari, W. N. A. W. (2015). An overview of the oil palm industry in Malaysia. *Renewable and Sustainable Energy Reviews, 50,* 1469–1484.

5. Loh, S. K., Lai, M. E., Ngatiman, M., Lim, W. S., Choo, Y. M., Zhang, Z., et al. (2013). Zero discharge treatment technology of palm oil mill effluent. *Journal of Oil Palm Research, 25*, 273–281.
6. Alengaram, U. J., Mohottige, N. H. W., Wu, C., Jumaat, M. Z., Poh, Y. S., & Wang, Z. (2016). Response of oil palm shell concrete slabs subjected to quasi-static and blast loads. *Construction and Building Materials, 116*, 391–402.
7. Nayaka, R. R., Alengaram, U. J., Jumaat, M. Z., Yusoff, S. B., & Alnahhal, M. F. (2018). High volume cement replacement by environmental friendly industrial by-product palm oil clinker powder in cement–lime masonry mortar. *Journal of Cleaner Production, 190*, 272–284.
8. Kanadasan, J., & Abdul Razak, H. (2015). Engineering and sustainability performance of self-compacting palm oil mill incinerated waste concrete. *Journal of Cleaner Production, 89*, 78–86.
9. Karim, J. M. R., Hashim, H., & Abdul Razak, H. (2016). Assessment of pozzolanic activity of palm oil clinker powder. *Construction and Building Materials, 127*, 335–343.
10. Ahmmad, R., Alengaram, U. J., Jumaat, M. Z., Sulong, N. H. R., Yusuf, M. O., & Rehman, M. A. (2017). Feasibility study on the use of high volume palm oil clinker waste in environmental friendly lightweight concrete. *Construction and Building Materials, 135*, 94–103.
11. Nayaka, R. R., Alengaram, U. J., Jumaat, M. Z., Yusoff, S. B., & Sumesh, M. (2019). Influence of palm oil clinker powder on the fresh and mechanical properties of masonry mortars. *IOP Conference Series: Materials Science and Engineering, 431*, 082002.
12. Alnahhal, M. F., Alengaram, U. J., Jumaat, M. Z., Abutaha, F., Alqedra, M. A., & Nayaka, R. R. (2018). Assessment on engineering properties and CO_2 emissions of recycled aggregate concrete incorporating waste products as supplements to Portland cement. *Journal of Cleaner Production, 203*, 822–835.
13. Kanadasan, J., & Razak, H. A. (2015). Utilization of palm oil clinker as cement replacement material. *Materials (Basel), 8*(12), 8817–8838.
14. Nayaka, R. R., Alengaram, U. J., Jumaat, M. Z., & Yusoff, S. B. (2018). Microstructural investigation and durability performance of high volume industrial by-products-based masonry mortars. *Construction and Building Materials, 189*, 906–923.
15. ASTM C270–14. (2014). *Standard specification for mortar for unit masonry*. West Conshohocken, PA: ASTM International.
16. Kanadasan, J., & Razak, H. A. (2014). Mix design for self-compacting palm oil clinker concrete based on particle packing. *Materials and Design, 56*, 9–19.
17. Karim, M. R., Hashim, H., Abdul Razak, H., & Yusoff, S. (2017). Characterization of palm oil clinker powder for utilization in cement-based applications. *Construction and Building Materials, 135*, 21–29.
18. ASTM C109, 109M-14. (2014). *Standard test method for compressive strength of hydraulic cement mortars (using 2-in. or [50-mm] cube specimens)*. West Conshohocken, PA: ASTM International.

Conformity of the Use of Space and Land Use in North Jakarta Administrative City

Galih Jati Utomo, Hafid Setiadi, and Rudy P. Tambunan

Abstract North Jakarta City as part of the Special Capital Region of Jakarta Province has a region characteristic typically of coastal areas with varied land use functions. Urban development that occurs along with population growth and development activities increases the demand for land, though the city is designed in a planological manner according to the type of designation and land use that has been previously determined. This condition leads to the occurrence of land use deviation from the North Jakarta Regional Spatial Plan (RTRW) and Detailed Spatial Plan (RDTR). This study aims to determine the pattern of changes in land use and utilization (spatial and temporal). The method used in this study was map analysis with overlay using geographic information system, qualitative data collection and quantitative descriptive method by identifying and conducting field survey observations. The results showed a change in the pattern of spatial utilization in 2008 until 2018. The land use that showed increase in North Jakarta City was industrial land and warehousing of 9.26%, whereas the most decreasing land use was occupancy/housing of 11.83%. Based on the results of the analysis, the land use that had conformity with the spatial plan in 2018 was 51.05%, 12.48% supported the spatial plan, 29.48% had no conformity with the spatial plan and 6.96% had not approved yet.

Keywords Spatial analysis · Spatial violation · Spatial planning · Spatial use · Land use

1 Introduction

A common problem faced by big cities in Indonesia is the high growth of urban population. This is due to natural population growth and urbanization factors. These

G. J. Utomo (✉) · R. P. Tambunan
Urban Development Studies, School of Strategic and Global Studies, University of Indonesia, Jakarta, Indonesia
e-mail: galihju@gmail.com

H. Setiadi
Department of Geography, Faculty of Mathematics and Natural Science, University of Indonesia, Jakarta, Indonesia

S. S. Emamian et al. (eds.), *ICSDEMS 2019*,
https://doi.org/10.1007/978-981-15-3765-3_2

two factors ultimately have an impact on the emergence of various problems in urban areas such as the lack of space for the needs of housing, offices, trade and services. The current urban development has an impact on increasing the need for space in urban areas and the high demand for land. Therefore, spatial planning is needed to regulate utilization patterns based on activities, types of activities, location functions, space quality and environmental aesthetics [1].

Based on the DKI Jakarta 2030 Spatial Planning (RTRW), in order to control the development of activities so as not to exceed the carrying capacity and accommodative capacity of the environment, one strategy is to prioritize the development of cities to the east, west and north and limit development to southward [2]. This is also in line with the vision of the North Jakarta Administrative City, namely to realize the North Jakarta Administrative City as a modern coastal city that is neatly organized, advanced, comfortable and prosperous and has a cultured society and a government-oriented public service. As a coastal city, North Jakarta has a strategic value in controlling space utilization in the DKI Jakarta Province.

Problems in the City of Jakarta related to the control of spatial use are land use and spatial use that do not comply with applicable laws and regulations. This is a factor that triggers the non-conformity of urban space utilization. Beneath that problem, DKI Jakarta Province still does not have any Regional Regulation on the Management Zoning of Small Coastal Areas and Islands (RZWP3K). The unresolved problem this spatial plan has been making the weak utilization control space, especially in coastal areas [3]. This issue also occurred in the North Jakarta Administrative City, and limited land and the high price of land in Jakarta and North Jakarta in particular caused an increase in the intensity of land use in each ownership.

This study is very important considering the great benefits obtained when the land use and spatial use can be identified. Through this study, it is expected to be known to what extent the conditions, dynamics and the extent of non-conformity that occurred in spatial planning in the North Jakarta Administrative City. Thus, evaluations can be carried out in the context of proposed changes or providing input on the revision of the Regional Spatial Plan (RTRW) through approaches that are in line with environmental conditions and problems. Based on the facts described above, the authors are interested to conduct a study entitled "Conformity of The Use of Space and Land Use in North Jakarta Administrative City."

According to Gallion and Eisner [4], changes in an area and part of a city are influenced by the geographical location of a city. There are several determinants in urban growth, such as transportation networks and relationships between cities and elements of the population. This is due to an increase in human activities in urban areas. The main urban activities that play an important role in urban development are trade activities, industrial activities and housing activities [5].

Land usability planning is often interchangeable with the term land use planning, because it basically has the same understanding. In some literatures, it is called land use planning. The difference between the two lies in the emphasis on space. According to Baja [6], land use implicitly contains the definition of space in it, because it is related to the use, arrangement or setting of the use both in the context of space and time (spatial and temporal). Meanwhile, Shirvani [7] mentions that each city design must pay attention to the existing design elements so that the city will have clear characteristics. There are eight elements that make up a city (mainly the city center), and one of them is land use. Allotment of land (land use) must pay attention or adjust to the problems in how to develop the city. The study conducted by Shirvani [7] also mentioned that zoning ordinance is a practical and useful control mechanism in urban design. The main emphasis lies on the three-dimensional problem, namely the harmony relationship between buildings and environmental quality. This is the basis for urban spatial planning.

Furthermore, Kivell [5] states that land use planning and zoning are instruments in spatial planning. These two instruments are generally detailed in the structure plan or in more detail in the area plan that shows where development can be permitted and in what form of building. The objectives of the area plan are limit the use of permitted space; guarantee the balance of land use for all activities in all strategic areas; avoid non-strategic land use.

In land use theory, land use is defined as land use regulation [8]. In land use, land can be interpreted as land that already has ownership and allocation status. Land is a limited natural resource as a necessity of life for every individual; its use needs to be regulated and planned. Land use is closely related to daily activities held on the plots of land. Thus, there are land uses for agriculture, industry, housing, education and so on. These land use patterns are regulated or planned in land use planning [9].

The theory of changes in an area and its surroundings as part of a wider urban area according to Gallion and Eisner [4] in the book, entitled The Urban Pattern, may cause distortions in land use changes. Based on Chapin and Kaiser's theory [10], urban development will always be associated with urban land use where there are three key systems that influence. The three systems include (a) the city activity system; (b) environmental system; (c) land development system. Meanwhile, land conformity is the level of conformity of a plot of land for a particular use. The conformity of the land can be assessed for current conditions (actual land conformity) or after improvements have been made (potential land conformity) [11].

2 Methods

This study uses a mix-method approach with a stepwise method, namely map analysis and quantitative and qualitative descriptive analyses by conducting in-depth interview with key informant. Changes in land use in the North Jakarta Administrative City area were obtained by analyzing digital maps by overlaying 2008 land use map and 2018 land use map as the secondary data from local government agencies. The results of the overlay produced a map of land use changes at the specified year spacing.

Then, to obtain the amount of land use non-conformity, it was obtained by classifying the forms of land use or classifications of space allocation on the RTRW which were determined based on the classification standards of the themes listed in the 2010–2030 RTRW Jakarta Spatial Pattern Map. The map used was a comparison of Detailed Spatial Plan map with land use map in 2018.

The amount of space utilization and the non-conformity to spatial planning were obtained from visual interpretation using geographic information systems (GISs). Geographic information system (GIS) is a computer-based system of work that has the ability to enter, manage, analyze and reactivate data that have spatial references for various purposes related to mapping and planning. The final result (output) in the form of geographical display is very useful for decision making [12].

3 Result and Discussions

3.1 Land Use Change

Based on the analysis results of the map of land use in 2008 (Fig. 1) and 2018 (Fig. 2) with interpretation of aerial photographs (satellite imagery) based on secondary data from the Department of Spatial Planning and Land Management of DKI Jakarta Province, there were 26 (twenty-six) classes or land use classification in North Jakarta Administrative City. To simplify land use classification, the authors simplified the classification into 6 (six) types, including a. facilities and means; b. occupancy/housing; c. industry and warehousing; d. offices, trade and services; e. open space, cemeteries, plantations, agriculture and vacant land and f. others. Based on an analysis of land use change, it was shown the pattern of land use in 2008 (Fig. 3)

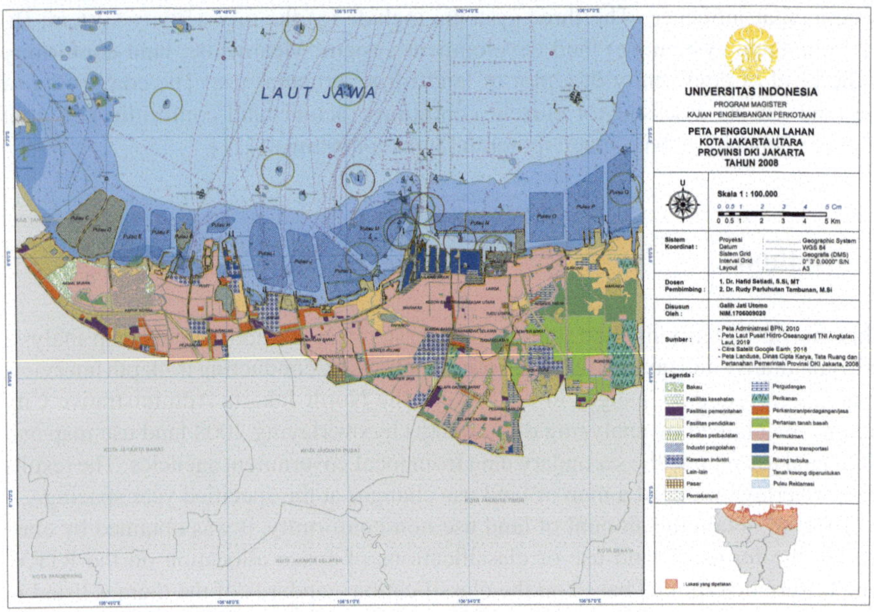

Fig. 1 Map of land use in North Jakarta Administrative City in 2008

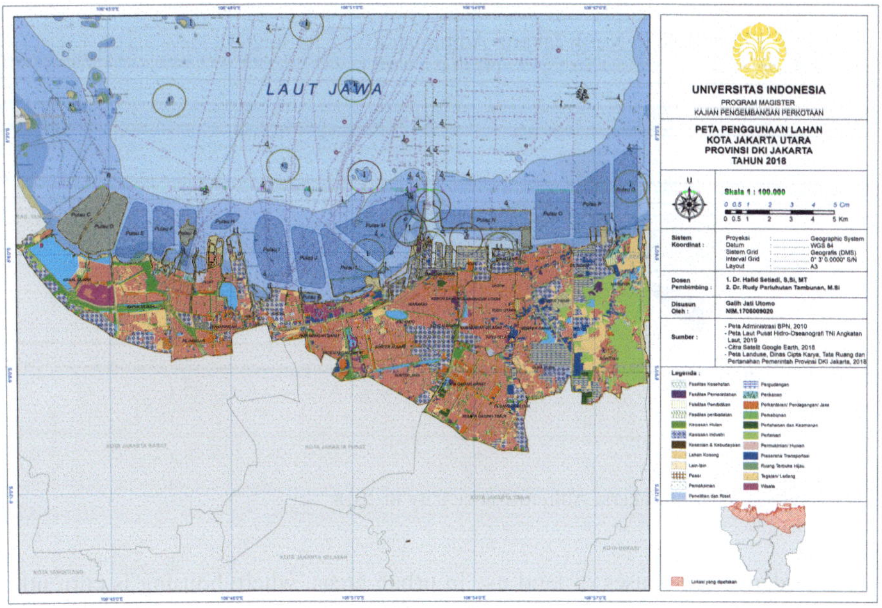

Fig. 2 Map of land use in North Jakarta Administrative City in 2018

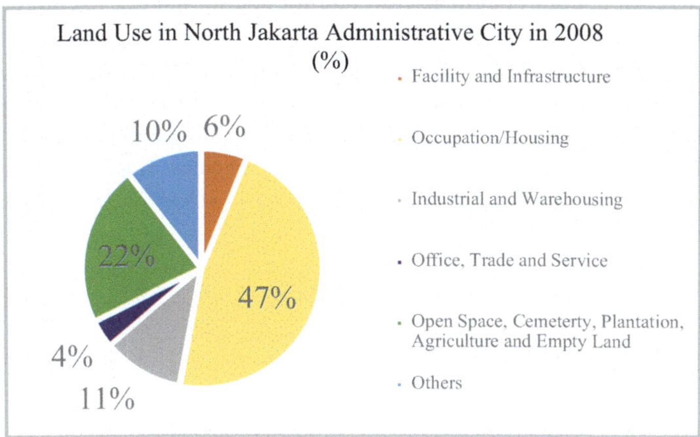

Fig. 3 Land use in North Jakarta Administrative City in 2008

and 2018 (Fig. 4) in North Jakarta City. The details for each subdistrict and land use changes at that time can be seen in Table 1.

Spatially, a subdistrict that has access to or is directly adjacent to a port and infrastructure in the form of accessibility to an activity center and an adequate road network has a high level of change. The results of the analysis showed that the use of residential land experienced a significant decline in the period of 2008–2018. This

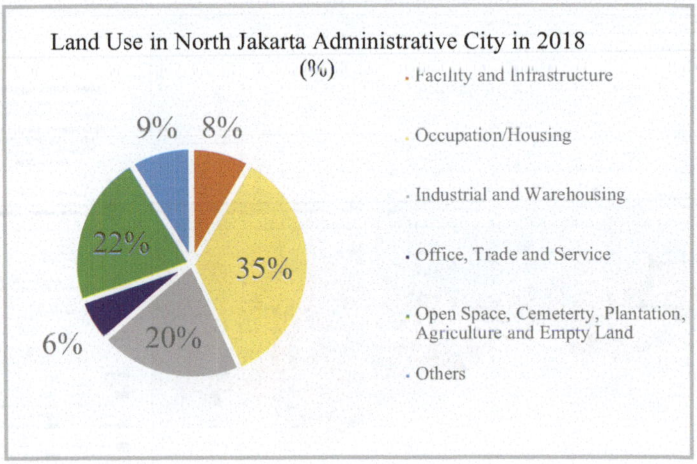

Fig. 4 Land use in North Jakarta Administrative City in 2018

is in accordance with cases of land use in urban areas, where housing is currently directed as vertical housing. In accordance with the DKI Jakarta Spatial Planning, achieving good density condition for Jakarta can be done through increasing land carrying capacity and intensifying land use through vertical development.

As a coastal area and to facilitate the need for space, plans to develop industrial estates and warehouses are indeed part of the spatial pattern plan in North Jakarta City. In addition, in the spatial pattern plan referred to in the cultivation area plan, a plan to develop industrial estates and warehouses was established to support the activities of Tanjung Priok Port. This can be seen from the increased use of space in the form of processing industries and warehousing services. Based on the results of the analysis, there was an increase of 9.26% (from 11.02 to 20.28%) for industrial and warehousing use.

3.2 *Land Use Conformity*

As a benchmark for conducting an overlay analysis, the matrix was first set as a tool for the land use conformity assessment. The matrix in this study was used to compare two factors, in which the column section was the spatial planning zoning and the row section was the classification of land use. The column and row show the strength of the relationship between items or nonlinear relationship.

The 2018 land use conformity matrix for Detailed Spatial Plan in North Jakarta City can be seen in Table 2. Matrix analysis referred to the approach adopted to Minister of Public Works Regulation No. 20 of 2011 concerning Guidelines for Preparation of Detailed Spatial Plan and District/City Zoning Ordinance [13]. The author limited the determination that did not refer to zoning ordinance where activities

Table 1 Land use in North Jakarta Administrative City in 2008 and 2018

Subdistrict region	Land use (in ha)											
	Facility and infrastructure		Occupation/housing		Industrial and warehousing		Office, trade and service		Open space, cemetery, plantation, agriculture, and empty land		Others	
	2008	2018	2008	2018	2008	2018	2008	2018	2008	2018	2008	2018
Penjaringan	132.18	140.77	1497.5	993.8	173.28	463.74	252.84	177.11	420.18	591.94	435.49	584.51
Pademangan	105.34	137.44	532.17	300.24	136.18	149.44	150.3	110.65	138.12	177.82	11.097	150.3
Tanjung Priok	186.08	131.87	1327	871.11	282.26	518.69	17.129	154.41	98.309	169.97	111.25	83.637
Koja	152.99	117.95	723.83	533.59	75.397	215.09	21.688	67.257	67.759	90.127	4.356	24.856
Kelapa Gading	134.73	132.43	843.99	646.35	203.58	144.71	9.7427	171.87	122.91	225.77	129.01	21.858
Cilincing	37.052	344.37	784.09	876.21	474.04	956.29	9.2821	54.328	1814.3	1365.1	581.55	175.44
Total	748.37	1004.8	5708.6	4221.3	1344.7	2448	460.99	735.62	2661.6	2620.7	1272.8	1040.6
(%)	6.13	8.32	46.80	34.97	11.02	20.28	3.78	6.09	21.82	21.71	10.43	8.62

Table 2 Analysis matrix of land use conformity with the Regional Spatial Plan of North Jakarta City

Detailed Spatial Planning (RDTR) / Land use	Mixed zone	Recreational zone	Industrial and warehousing zone	Road zone	Green zone	Protected zone	Public service and social zone	Regional government zone	National government zone	Office, trade and service zone	Office, trade and service zone Low building basic coefficient	Village housing zone	Housing zone Low building basic coefficient	Housing zone Moderate-high building basic coefficient	Vertical housing zone	Vertical housing zone low building basic coefficient	Cemetery zone	Reclamation zone	City park zone /environment	Blue open zone
Facility and infrastructure																				
1. Health facility	M	TS	TS	TS	TS	TS	S	S	S	S	S	M	M	M	M	M	TS	TS	TS	TS
2. Government facility	S	TS	S	TS	TS	TS	S	S	S	M	M	TS	TS	TS	TS	TS	TS	TS	TS	TS
3. Education facility	S	TS	TS	TS	TS	TS	S	S	S	M	M	M	M	M	M	M	TS	TS	TS	TS
4. Religious facility	M	TS	M	TS	TS	TS	S	M	M	M	M	M	M	M	M	M	TS	TS	TS	TS
5. Telecommunication facility	M	TS	M	TS	TS	TS	S	M	M	M	M	M	M	M	M	M	TS	TS	TS	TS
6. Art and culture facility	M	TS	TS	TS	TS	TS	S	M	M	M	M	M	M	M	M	M	TS	TS	TS	TS
7. Market	M	TS	TS	TS	TS	TS	S	S	S	S	S	TS	TS	TS	TS	TS	TS	TS	TS	TS
8. Public service	M	TS	M	TS	TS	TS	S	M	M	M	M	M	M	M	M	M	TS	TS	TS	TS
9. Study and research	M	TS	TS	TS	TS	TS	S	M	M	M	M	TS	TS	TS	TS	TS	TS	TS	TS	TS
Residential/housing																				
1. Residential, housing, settlement	S	TS	TS	TS	TS	TS	M	TS	TS	M	M	S	S	S	S	S	TS	TS	TS	TS
Industrial and warehousing																				
1. Industrial	TS	TS	S	TS	TS	TS	TS	TS	TS	M	M	TS	TS	TS	TS	TS	TS	TS	TS	TS
2. Warehousing	TS	TS	S	TS	TS	TS	TS	TS	TS	M	M	TS	TS	TS	TS	TS	TS	TS	TS	TS
Office, trade and service																				
1. Office, trade and service	S	TS	M	TS	TS	TS	TS	TS	TS	S	S	TS	TS	TS	TS	TS	TS	TS	TS	TS
Open space																				
Forest area	M	S	M	TS	S	S	M	M	M	M	M	M	M	M	M	M	M	TS	S	TS
1. Empty land	M	M	TS	TS	M	M	M	M	M	M	M	M	M	M	M	M	M	TS	S	TS
2. Cemetery	TS	TS	TS	TS	TS	TS	TS	TS	TS	TS	TS	TS	TS	TS	TS	TS	S	TS	M	TS
3. Plantation	TS	M	TS	TS	TS	TS	TS	TS	TS	TS	TS	TS	TS	TS	TS	TS	TS	TS	S	TS
4. Agriculture	TS	TS	TS	TS	TS	TS	TS	TS	TS	TS	TS	TS	TS	TS	TS	TS	TS	TS	S	TS
5. Open space	M	S	M	TS	S	M	M	M	M	M	M	M	M	M	M	M	M	TS	S	TS
6. Agricultural field	TS	TS	M	TS	S	M	M	M	M	M	M	M	M	M	M	M	M	TS	S	TS
Other allocation																				
1. Waters	TS	TS	TS	TS	TS	S	TS	TS	TS	TS	TS	TS	TS	TS	TS	TS	TS	TS	M	S
2. Fishery	TS	TS	TS	TS	TS	TS	TS	TS	TS	M	M	TS	TS	TS	TS	TS	TS	TS	TS	S
3. Defense and security	TS	TS	S	TS	TS	TS	S	S	S	TS	TS	TS	TS	TS	TS	TS	TS	TS	TS	TS
4. Transportation	TS	TS	M	S	TS	TS	S	TS	TS	TS	TS	TS	TS	TS	TS	TS	TS	TS	TS	TS
5. Tourism	TS	S	TS	TS	TS	TS	M	TS	TS	M	M	TS	TS	TS	TS	TS	TS	TS	S	TS
6. Others	TS	TS	TS	S	S	TS	TS	TS	TS	TS	TS	TS	TS	TS	TS	TS	TS	BS	M	S

Information		
S	:	Appropriate
M	:	Supportive
TS	:	Inappropriate
BS	:	Not Approved Yet

were conditional and limited to the ITBX table Regional Regulation No. 1 of 2014 concerning Detailed Spatial Plan and zoning ordinance in DKI Jakarta Province [14]; however, the determination of land use conformity was close to that.

From the results of the overlapping analysis using 2018 land use map with spatial planning, it can be seen that there was "inappropriate" land use with "supportive" land use to the North Jakarta Administrative City plan. In conducting the analysis, the authors calculated the appropriate, not appropriate and supportive land area. Conformity between the types of land use for the functions planned in the spatial planning in the North Jakarta Administrative City was based on four criteria as follows:

1. **Appropriate (S)**, if land use has been fully was in accordance with regional functions in the Detailed Spatial Plan (RDTR).

2. **Inappropriate (TS)**, if land use was not in accordance with regional functions in the Detailed Spatial Plan (RDTR).
3. **Supportive (M)**, if land use was not in accordance with regional functions but its existence could be maintained or adjusted as long as it did not interfere and could contribute to the realization of regional functions as stipulated in the Detailed Spatial Plan (RDTR).
4. **Not Approved Yet (BS)**, if the land allocation conditions had not been established or had not been approved in the Detailed Spatial Plan (RDTR).

Especially for reclamation islands in this study were included in the classification of other land allocation with a study matrix in the form of "plans that have not been approved." This is because the reclamation islands do not yet have zoning and regulations that are set as the legal basis in the management of coastal areas and small islands located in Jakarta Bay (Regional Regulation on the Management Zoning of Small Coastal Areas and Islands [RZWP3K]).

The results of the conformity overlay between spatial pattern plan contained in the Regional Spatial Plan dan Detailed Spatial Plan with the land use map in 2018 can be seen in Table 3 and the conformity map is presented in Fig. 5.

In calculating the conformity amount, this study used the Minister of Agrarian and Spatial Planning (ATR)/Head of National Land Agency (BPN) Regulation No. 6 of 2017 concerning Procedures for Reviewing the Regional Spatial Plan to calculate the percentage of the area of each planned area [15].

$$\frac{A}{x} \times 100\% = a\% \tag{1}$$

a Land use conformity value
A Existing physical shape
x Area according to spatial plan.

The percentage of conformity between spatial pattern plan contained in the Reginala Spatial Plan dan Detailed Spatial Plan with land use in 2018 can be seen in Fig. 6 while the conformity map is presented in Fig. 5 as the result of overlaying the two. Based on Figs. 5, 6 and 7, it was known that 7650.24 ha of land use in 2018 or 51.05% of the total area of North Jakarta City were appropriate based on the spatial pattern plan, whereas 1869.85 ha (12.48%) were supportive spatial pattern plan, 4422.42 ha (29.51%) were not appropriate and 1042.87 ha (6.96%) of the plans had not yet been approved yet. The region with the highest conformity value was Kelapa Gading District of 64.16% and the lowest was Cilincing District of 35.32%.

Then, based on the Minister of Agrarian Affairs and Spatial Planning (ATR)/Head of National Land Agency (BPN) No. 6 of 2017 concerning Procedures for Reviewing Regional Spatial Planning, the calculation of utilization conformity (space pattern) could be taken as follows:

$$\frac{7650.24\ \text{ha}}{14{,}985.38\ \text{ha}} \times 100\% = 51.05\% \tag{2}$$

Table 3 Percentage of land use conformity in 2018 with North Jakarta City Detailed Spatial Plan based on subdistrict

Subdistrict region	Appropriate		Inappropriate		Supportive		Not approved yet		Total	
	Hectare	Percentage	Hectare	Percentage	Hectare	Percentage	Hectare	Percentage	Hectare	Percentage
Penjaringan	2303.32	54.43	416.77	9.85	826.04	19.52	685.28	16.20	4231.42	100.00
Pademangan	666.40	49.98	278.25	20.87	336.86	25.26	51.88	3.89	1333.38	100.00
Tanjung Priok	1423.24	63.49	330.60	14.75	485.73	21.67	2.12	0.09	2241.70	100.00
Koja	661.04	57.80	116.77	10.21	365.20	31.94	0.57	0.05	1143.58	100.00
Kelapa Gading	1033.90	64.16	187.57	11.64	389.91	24.20	0.00	0.00	1611.38	100.00
Cilincing	1562.34	35.32	539.89	12.20	2018.68	45.63	303.02	6.85	4423.93	100.00
Total	7650.24	51.05	1869.85	12.48	4422.42	29.51	1042.87	6.96	14,985.38	100.00

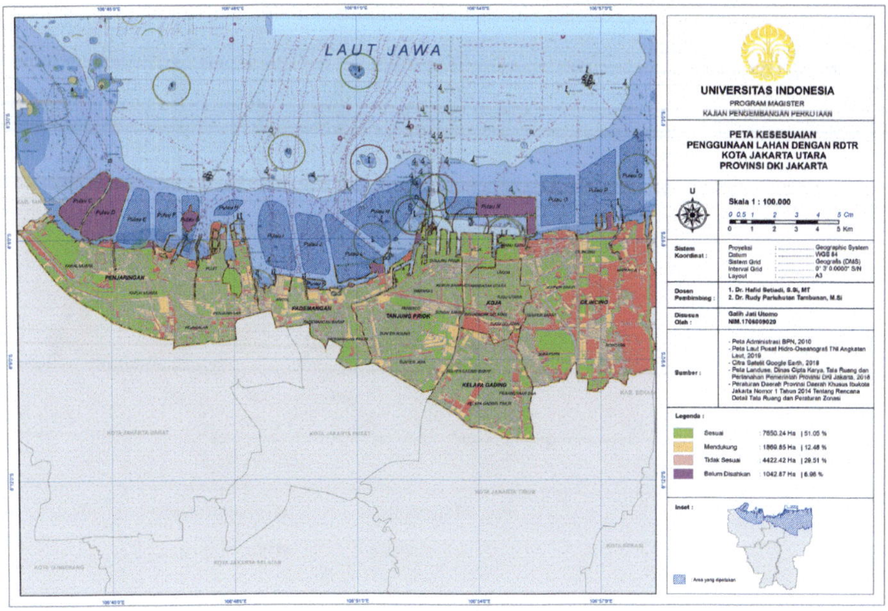

Fig. 5 Map of land use conformity in 2018 with North Jakarta City Detailed Spatial Plan, North Jakarta City

Fig. 6 Percentage of land use conformity in 2018 with North Jakarta City Detailed Spatial Plan

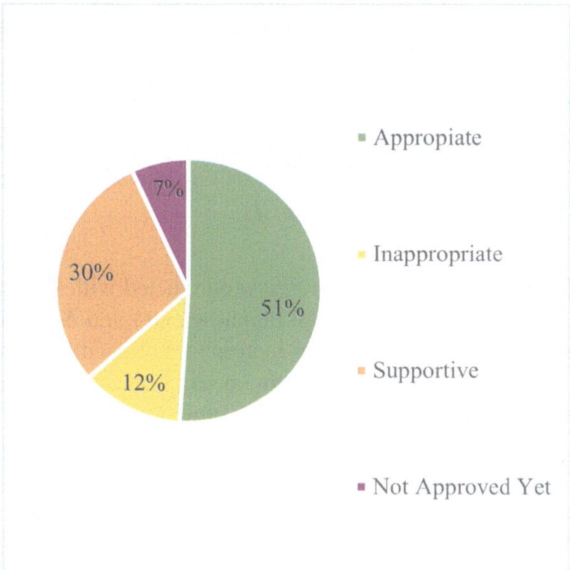

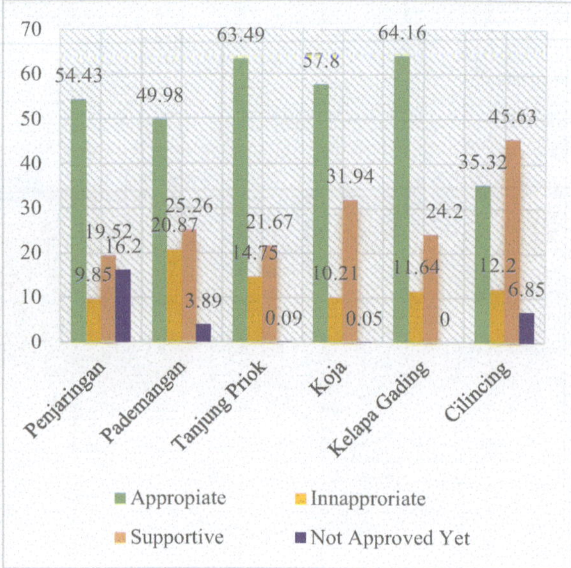

Fig. 7 Percentage of land use conformity in 2018 with North Jakarta City Detailed Spatial Plan based on subdistrict

With a conformity value of 51.05% and viewed from the aspect of realization of the type and amount of spatial utilization implementation, North Jakarta City could be said to have a moderate category. This was due to the realization value of the space utilization implementation between 51.05 up and 62.53% if the support category was also included in the calculation value.

3.3 Factors That Influence and Impact and Impact of Land Use Non-conformity

Based on the result of interview conducted with key informant, namely the Head of Cipta Karya Office, Spatial Planning and Land Administration of North Jakarta City and field observations, the factors that allegedly influenced the occurrence of land use non-conformity in North Jakarta City included land/space ownership system, economic benefits and influence of neighborhood unit.

Coastal areas have a vulnerability to higher land change and development. The implication as a consequence of land use non-conformity toward spatial planning may cause threats to the carrying capacity and sustainability of life in the future. Problems that arise as a result of the spatial utilization non-conformity that coincide with the unavailability of environmental infrastructure in North Jakarta City resulted in the emergence of irregular housing, traffic congestion, natural vulnerabilities (floods), damage to environmental quality, reduced public space and inability to implement government development programs and so on.

4 Conclusions

Changes in land use utilization in the North Jakarta Administrative City occurred in several land uses. Land with increased utilization in North Jakarta between 2008 and 2018 was industrial and warehousing of 9.26%. Meanwhile, the land with the most decreased utilization was residential/housing of 11.83%. The increase in changes in land use is in accordance with the direction of the DKI Jakarta Spatial Plan, which states that the spatial pattern of North Jakarta City is to provide warehousing and industrial facilities to support trade activities and services to support the activities of Tanjung Priok Port.

Compliance between existing land use in 2018 with Detailed Spatial Plan in North Jakarta Administrative City showed a value of 51.05 up to 62.53% which was included in support category. This figure was categorized in moderate level in the evaluation of spatial use conformity based on Minister of Agrarian Affairs and Spatial Planning (ATR)/Head of National Land Agency (BPN) No. 6 of 2017 concerning Procedures for Reviewing Regional Spatial Plan.

The implications of land use non-conformity in North Jakarta City included the emergence of irregular housing, traffic congestion, natural vulnerabilities such as floods, damage to environmental quality, reduced public space and inability to implement government development programs and so on. And approach to overcome these problems is by reviewing spatial plans, increasing the role of the government in monitoring and controlling spatial use and making priority programs and areas with a tight control system.

Acknowledgements This study is part of the Master's Indexed International Publication Study (PITMA B) 2019–2020 of the University of Indonesia with the theme of "Practices of Planning and Conflict of Spatial Use in Urban Development," contract number NKB-1042/UN2.R3.1/HKP.05.00/2019.

References

1. Sugandhy, A. (1994, February 2). Operationalization of spatial planning and development trilogy. *PRISMA Economic and Social Study Magazine*.
2. Special Capital Region of Jakarta Province. (2012). *Regional regulation no. 1 of 2012 concerning Regional Spatial Planning in 2010–2030*.
3. Mungkasa, O. M. (2015). *Jakarta: Problem and solution*. Paper presented at Open Selection Process for DKI Jakarta Deputy Governor for Spatial Planning and Environment.
4. Gallion, A. B., & Eisner, S. (1986). *The urban pattern: City planning and design*. New York: Van Nostrand.
5. Kivell, P. (1993). *Land and the city: Pattern and process of urban changes*. London: Routledge.
6. Baja, S. (2012). *Land use planning in regional development—Spatial approaches and their applications*. Yogyakarta: Andi.
7. Shirvani, H. (1985). *The urban design process*. New York: Van Nostrand Reinhold Company Inc.
8. Jayadinata, J. (1996). *Land use in rural, urban and regional planning*. Bandung: ITB Publisher.

9. Soefaat. (1999). *Functional relations of civil engineering with urban and Regional Spatial Planning*. Jakarta, PT: Mediatama Saptakarya.
10. Chapin, F. S., & Kaiser, J. (1979). *Urban land use planning*. Chicago: University of Chicago Press.
11. Sitorus, S. R. (1998). *Evaluation of land resources* (3rd ed.). Bandung: TARSITO Publisher.
12. Muta'ali, L. (2013). *Regional and city spatial organization (normative-technical review)*. Yogyakarta: Publisher Agency of Faculty of Geography (BPFG), Gajah Mada University.
13. Republic of Indonesia. (2011). *The Minister of Public Works regulation no. 20/PRT/M/2011 concerning guidelines for preparation of detailed spatial planning (RDTR) and district/city zoning ordinance.*
14. Special Capital Region of Jakarta Province. (2014). *Regional regulation number 1 of 2014 concerning detailed spatial planning and zoning ordinance.*
15. Republic of Indonesia. (2017). *The Minister of Agrarian Affairs and spatial planning/head of national land agency (BPN) regulation no. 6 of 2017 concerning procedures for reviewing Regional Spatial Planning.*

Factors Affecting the Perception of the Surrounding Community in Maintaining the Sustainability of the Setu Babakan Betawi Cultural Village in South Jakarta

Dwinanto Suberlian and Chotib

Abstract The Setu Babakan Betawi cultural village is an area that has been established by the Government for Betawi culture preservation since 2000. The purpose of this study is to analyze the current landuse and at the time of its establishment as a Betawi socio-cultural preservation area, analyze the attitudes and expectations of the surrounding community toward the existence and sustainability, and analyze the variables that influence people's perceptions of area. A total of 345 questionnaires were distributed using a simple random sampling method. Descriptive analysis is used to find out the categories of respondents' statements from each variable indicator. While quantitative analysis using structural equation modeling (SEM) analysis techniques aims to measure how significant the relationship between variables and indicators [1], there was a change in the land use from 2005 to 2019, namely the reduction of green open space and the increase in residential land area. Community perceptions are shown in a statement supporting and agreeing that the home environment of the residence is included in the area [2]. The hopes and desires of the people around Betawi Village are that the community can continue to be in the area and be involved in activities within the area. Economic factors are the most important variables for the community to maintain sustainability because the community feels very economically helped by opening small businesses in the region. Social factors also play an important role in maintaining the sustainability. Based on the results of path analysis from structural equation models, it is known that public perception is significantly influenced by economic factors and social factors, while environmental factors do not significantly influence people's perceptions.

Keywords Betawi cultural village · Perception · Maintain · Existance · Sustainability · Factors

D. Suberlian · Chotib (✉)
Urban Studies, School of Strategic and Global Studies, Universitas Indonesia, Depok, Indonesia
e-mail: chotib@hotmail.com

S. S. Emamian et al. (eds.), *ICSDEMS 2019*,
https://doi.org/10.1007/978-981-15-3765-3_3

1 Introduction

The establishment of the Betawi cultural village area is not widely known by new people came to Jakarta. Not all surrounding communities support cultural tourism programs due to inappropriate understanding such as: lack of knowledge and understanding of the surrounding community, regional support facilities that have not been much realized, and the involvement of the surrounding community, which is less involved in regional planning [3]. Furthermore, the research question in this research is How is the change at the time of determination in the Betawi Village region compared with the current conditions? What is the perception of the surrounding community about the change? What are the efforts of the community toward the existence and sustainability of Betawi Village? What efforts should be made by the community in order to maintain the existence and sustainability of Betawi Village.

2 Methods

This research aims to test the theory by using a mix of quantitative and qualitative methods in one period of time to produce an overall information for the interpretation of results (Table 1).

In order to get a representative sample, one person represents each house who can be the head of the family or one of the household members who are older adults who live in the area. The sample size of the population uses the simple random sampling method with the aim that all members of the population have the same opportunity to be chosen regardless of the strata in the population. 345 samples of this study were calculated using the Slovin formula.

3 Result and Discussions

3.1 Respondent Profile

Respondents were people were settled in this location before the Betawi Village location determination was set by the DKI Government in 2000. If compared with the main occupational profile of respondents who generally are outside the area, this would indicate that the mobility of people coming out and entering this region was low. Accessibility of this business will be a potential key and key to support the region and an effort that can be made if community perception does not support the existence and sustainability of the Betawi Village area.

Table 1 Operational conceptual variables

Variable	Indicator	Size
Population profile	1. Age 2. Gender 3. Duration of education 4. Own house 5. Length of stay 6. Number of family members 7. Betawi ethnic native population 8. The type of work is directly related to the Betawi cultural village activities 9. The main source of income for the house dealing directly with a ladder attraction	Year Nominal Year Yes/No Year Person Yes/No Yes/No Yes/No
Community perception	1. Determination of living environment as a Betawi cultural area 2. The purpose of the Betawi Village area is to revive Betawi culture 3. The neighborhood is better and more comfortable 4. Plans for additional Betawi Village area 5. There is a characteristic of Betawi architecture in the house of residence 6. Open living environment for tourist visitors 7. The living environment is more organized and neat 8. Management of residential environment by the DKI Jakarta Regional Government 9. The neighborhood is a tourist attraction 10. Special locations for trading activities will be developed	Likert scale 1–5 where: 1 = very not agree 5 = strongly agree
Social	1. The existence of social relations/social ties 2. Close relationship with neighbors 3. Knowing each other and communication 4. Neighbors help each other 5. Desire and behave together 6. Participation [4]	Likert scale 1–5 where: 1 = very not agree 5 = strongly agree
Environment	1. Availability of clean water 2. Environmental quality is clean from air, water, and sound pollution 3. Green open space 4. Availability of education, health, social and public facilities 5. Availability of public open space	Likert scale 1–5 where: 1 = very not agree 5 = strongly agree
Economy	1. There is a chance to work 2. There are people/institutions that want to help finance 3. Ease of getting a loan 4. Selling, buying and renting a house to live in [5] 5. Facilities for trading/services	Likert scale 1–5 Where: 1 = very not agree 5 = strongly agree

Table 2 Land use existing and plan area 2005–2019

Land use	2005		Spatial detail plan 2010–2030		2019[a]	
	M2	%	M2	%	M2	%
Residential	1312888	46.89	1479123	51.18	1450346	51.64
Mix (commercial and residential)	–	–	–	–	114491	4.08
Tourism service	–	–	–	–	29712	1.06
Public facilities	124231	4.44	343246	11.88	157076	5.59
Open green space	1254547	44.80	658325	22.78	764359	27.22
Open blue space	108461	3.87	409303	14.16	292492	10.41
Total	2800128	100	2890000	100	2808479	100

[a]Interpreted May, 2019

3.2 *Land Use Existing*

The utilizations of area space that was carried out at the time of the study mostly were residential and water open space. In addition, there are also mixed space uses such as shop houses, green open spaces (funerals, parks, agricultural land) and public facilities (schools, hospitals, health centers). Mixed functions (commercial and residential) are along the path of the main road of the area, namely M. Kahfi II Street with residential functions in the area behind it. The need for the development of new housing due to the development of the city using or taking up green open space means the reduced potential of land for green land (parks, plantations). The capacity for residential and mixed functions has exceeded the limit, and it is recommended that the government should immediately provide a policy of incentives and disincentives for landowners (communities) to maintain the existing land and not to change the function of their land from designated open land (gardens) to built land (Table 2).

4 Conclusions

Based on the relationship between variables that shape people's perceptions, the network variable is formed from neighboring life in the neighborhood (family). This shows that the statement of the respondent still believing and reintroducing the life values of the Betawi community in the past is again applied to the Betawi Village environment (Table 3).

The expected reciprocity is to get help from neighbors when needed. So, if you want to make people's perceptions to support the existence of the Betawi Village area, values must be sought to instill a passion for helping and giving importance to others. This will give birth to reciprocal value. To increase community participation in activities in a residential area, it is recommended to approach the local community

Table 3 Evaluation model measurement phase 2 (second-order outer model)

Var. latent	Variable manifest	Loading factor	Rule of thumb	Value	Rank
Social	Network	0.926	0.500	Valid	1
	Mutual	0.880	0.500	Valid	2
	Participation	0.639	0.500	Valid	3

through the potential characteristics of the local community both from the productive age, the level of education, the similarity of ethnic, and the length of stay in the region (Table 4).

To build the public perception that maintains the existence and sustainability of the area from the economic variable, it is recommended to improve facilities for trade/services. This is an effort to provide positive aspects to build people's perceptions. Economic aspects are the most significant thing in building people's perceptions to maintain Betawi Village sustainability. In the community perception variable, to strive for the existence and sustainability of the Betawi Village area, the effort that must be built is a combination of economic, social, and environmental variables. The best indicator in forming environmental variables is the quality of the environment clean from air, water, and sound pollutions. From a long period of stay of more than 10 years to more than 30 years, local people want the conditions of their living environment to be clean from water, air, and sound pollutions [6]. This shows that the community recalls the environmental conditions which at the beginning in the residential environment were still beautiful and clean.

The limited land and the rapid development of the city of Jakarta has encouraged rapid changes in the use of space from 2005 to 2019 from green open spaces to built land in the Betawi Village area. Changes in land ownership to migrants have become one of the problems in maintaining regional preservation because there will be a change in values and elimination of the characteristics of Betawi culture. The constraints on the lack of land in Betawi Village owned by the government cause changes in the use of non-built land into residential or home land due to the rapid development of the city, which can lead to reduce the image of the area as a Betawi cultural preservation area. The development of the Betawi Village area is still far from the expectations set out in the regional regulation concerning the arrangement

Table 4 Results of estimates and testing of hypotheses

Variable			Coefficient of path	Resulting test		Score
Independent	→	Dependent		CR	*p*-value	
Social	→	Community perception	0.237	3.249	0.001	Significant
Environment	→	Community perception	0.054	0.912	0.362	Not significant
Economic	→	Community perception	0.534	6.582	0.000	Significant

of the Betawi Village area. From the designated area, dense settlements have been developed, and there are difficulties in realizing the Betawi Village development master plan outside the core zone of the Situ Babakan area because the government should have immediately decided to allocate a new land to the regional development plan.

Community perceptions of Betawi Village shown in the statement of hopes and desires from the people around Betawi Village are that the community can continue to be in the area and to be involved in activities within the Betawi Village area. Economic factors are the most important variables for the community to maintain Betawi Village sustainability, because the community feels very economically helped by opening businesses in the region Betawi Village. Social factors also play an important role in maintaining the sustainability of Betawi Village. The most important dimension of the social variable is the network or social relations based on trust, reciprocity (reciprocal), and the existence of community participation in activities.

Acknowledgements This article is part of Grant of International Publications for Students Final Project of Universitas Indonesia: Sustainable Cities and Communities.

References

1. Chotimah, K, Mutiara. (2007). *Participation of Local Betawi Residents in the Context of the Sustainability of the Betawi Cultural Village program.* Thesis, Urban Development Studies, University of Indonesia, Jakarta.
2. Mulyana, D. (2005). *An Introduction to Theory of Communication.* Bandung: Remaja Rosdakarya.
3. Cook, B., & Khotani, U. (2001). *Participation The New Tyranny.* New York: Zed Books Ltd.
4. Winarsih, R., & Septi, A. (2005). *Management of Service.* Yogyakarta: Pustaka Belajar.
5. Cruddas, S. (2013). An Introduction to Structural Equation Modelling for Emergency Services and Disaster Research. *International Journal of Emergency Services, 2*(2), 131–140.
6. Efendi, U. (1984). *An Introduction to Psycology.* Bandung: Angkasa.
7. Sirgy, M. J. (2012). *The Psychology of Quality of Life: Hedonic Well-Being, 303 Life Satisfaction, and Eudaimonia* (Social Indikators Research Series 50).

Analysis of Pedestrian Movements Due to the Operations of Cibubur LRT Station

Wiby Adhitya Prayoga and Jachrizal Sumabrata

Abstract As a result of the construction of the Cibubur LRT Station, new commuters will come from and toward to the LRT station point. This has a positive impact because there is an increase in users of public transportation. However, this also had to cause new problems caused by the increasing volume of prospective passengers. This study aims to analyze the movement of walking and several scenarios for pedestrian access from or to the LRT station's central point to supporting facilities around the area and find the best solution to improve service levels. The station road access models are created using PTV VISSIM 11 software. A validation test is needed to determine whether the model is acceptable or not by comparing the results of the model and actual conditions in the field. Service level analysis (LOS) uses HCM 2000 as a reference standard.

Keywords Pedestrian · PTV Vissim 11 · Cibubur LRT Station · LOS

1 Introduction

1.1 Background

In this century era, the need for efficient public transportation is certainly a very urgent need. These situations are referred to as the condition of the highway is no longer available to accommodate the number of daily commuters, especially for private vehicles such as cars and motorbikes to work in Jakarta from Bodetabek cities. The concept of transit-oriented development (TOD) has begun to be planned massively in a long-term masterplan to facilitate the mobility of community activities. However, due to the condition of the satellite cities that are getting more and more crowded, new urban areas will emerge, such as in the Cibubur, Sentul, Cileungsi, and so on areas that have not been served by mass transportation. This is certainly the government's

W. A. Prayoga (✉) · J. Sumabrata
Civil Engineering Department, Faculty of Engineering, Universitas Indonesia, Kampus Baru UI, Depok 16424, Indonesia
e-mail: rjs@eng.ui.ac.id

S. S. Emamian et al. (eds.), *ICSDEMS 2019*,
https://doi.org/10.1007/978-981-15-3765-3_4

focus in building new infrastructures such as light rapid transit (LRT) along with supporting infrastructures such as in the Cibubur area and its surroundings.

1.2 Objective

The purpose of this study is to conduct modeling of the movement plan and direction of prospective passengers regarding the planned construction of the Cibubur LRT station area which will be implemented later, using the modeling application, PTV VISSIM 11.

2 Theory

2.1 LRT Cibubur Station

This Cibubur LRT Station will be built on the Kwartir Nasional (Kwarnas) land. Covering an area of 19 ha, later in this area, a terminus station will be built from the light rapid transit service (LRT) from Cibubur–Cawang line for 14.5 km and will be operational by the end of 2021. Later in and around the station, various kinds of public interactive spaces will be built, such as meeting rooms/ballrooms, food stands, and modern shopping areas to provide facilities for the prospective LRT passengers. In addition, of course, pedestrian access and parking lots will be built that are large enough to facilitate prospective passengers.

2.2 Model Analysis

LOS is an analytic concept related to passenger comfort and safety, which was developed in The Highway Capacity Manual 1965. It consists of six classes. LOS A is for roads where pedestrians can walk on a free-flow speed, and can walk there with no obstacle from people walking past it. While on LOS F, the density of pedestrians has reached a critical point, and pedestrians sometimes have to stop because they are disturbed by other pedestrians (Table 1).

2.3 Design Walkway Concept

According to the Transport Capacity and Quality of Service Manual [1], the purpose of the walkway concept is to determine the effective width, where this effective width

Table 1 Relationship between LOS and pedestrian flow

LOS	Space modulus M (m^2/p)	Unit flow rate v (p/min/m)
A	>5.6	<17
B	>3.7–5.6	>17–23
C	>2.2–3.7	>23–33
D	>1.4–2.2	>33–50
E	>0.7–1.4	>50–77
F	<0.7	Uncertain

Source HCM 2000 [5]

is the width provided for walking. According to PUPR [2], the effective width of the walkway can be calculated using the equation:

$$\mathrm{We} = \frac{v}{35} + N$$

whereas

We = Effective width (m)

v = Unit flow rate (p/min/m)

N = Additional width ($N = 1.5$ m for high number of pedestrians, $N = 1.0$ for medium number of pedestrians, and $N = 0.5$ m for low number of pedestrians on sidewalks).

3 Methods

This research was conducted to analyze the preliminary design of the construction of the Cibubur LRT Station area, particularly regarding access to pedestrian/pedestrian lines from and to the Cibubur LRT Station toward passenger destinations. The final stage of this research is to analyze whether the passenger/pedestrian access plan is appropriate and able to make prospective passengers comfortable and safe when the LRT is operated by the end of 2021. In addition, if the existing access scenario is not feasible, it can propose effective and efficient pedestrian access, by changing some of the variables used or by changing several alternative access station scenarios. The results of data processing are then processed so that they get a conclusion using a comparison of LOS analysis in every measurement area.

Fig. 1 Pedestrian route decision layout

4 Results

4.1 Research Variable

4.1.1 Design Plan Layout

According to the plan, the sidewalk is 2 m wide. This is the following determination of the network of pedestrian routes at the Cibubur LRT station. The division of the section area for measurement area analysis is summarized in the modeling layout (Fig. 1).

4.1.2 Modal Split

After defining the pedestrian routes, the modeling of the modal split variable will be divided according to the mode to be used and the area of origin/destination of the passenger. Broadly speaking, it can be summarized in Tables 2 and 3.

4.1.3 Passengers Volume (Number of Pedestrians)

Getting the number of passenger volumes to be entered into the PTV Vissim 11 model depends on several variables used, for example, The headway LRT (5 and 10 min), train occupancy factors (5 and 75%), and modal split variables, The total number of passengers per hour obtained from the calculation as follows:

Table 2 Modal split

Scenario model	Origin/destination passengers	
	Public transportation (%)	Private vehicle (%)
1	10	90
2	25	75
3	75	25
4	90	10

Table 3 Divided of modal split

Scenario models	Origin/destination passengers modal				
	Public transportation			Private vehicle	
	A1	A2	A3	A4	A5 and A6
All Scenario	40%	40%	20%	60%	40%

$$V = V_m \times n \times N \times \text{Occ};\ N = \frac{60}{H}$$

whereas

V = Number of passengers (p/h)

V_m = Max number of passengers in *Rush Load* per cars unit (p/cars)

n = Number of the cars on one train set (cars)

N = The frequency of trains arriving and departing in one hour (train/h)

Occ = Occupancy factors (50 or 75% of the total passenger capacity that can be transported in a series of LRT trains).

4.2 *Processing Data Input to the PTV Vissim 11*

As a result of several research variables that have been done processed before, the results of the input number of passengers that will arrive and depart from the Cibubur LRT station are then run in each simulation to get the desired results. The results are summarized in Table 4. We assumed the amount of departures passengers input are 50% from the arrival passengers input during the time of simulation (in the evening).

4.3 *Initial Design Plan Scenario (S0)*

The first simulation plan is the initial plan that will illustrate the possibility of the condition of the plan to be built according to the design from the contractor data. As planned, the width of the sidewalk will be 2 m wide on each pavement. In this

Table 4 Numbers of the arrival passenger input to PTV Vissim 11 at Cibubur LRT station (devided by Modal Split Percentages)

Model	Headway (min)	Occ (%)	Passengers (h)	Simulation 1							Simulation 2						
				To public transport				To private vehicle			To public transport				To private vehicle		
				10%				90%			90%				10%		
				A1	A2	A3	Total	A4	A51 and A52	Total	A1	A2	A3	Total	A4	A51 and A52	Total
				40%	40%	20%		60%	40%		40%	40%	20%		60%	40%	
2	5	75	10,800	432	432	216	1080	5832	3888	9720	3888	3888	1944	9720	648	432	1080
3	10	50	3600	144	144	72	360	1944	1296	3240	1296	1296	648	3240	216	144	360

Fig. 2 Modeling of running simulation from Cibubur Junction direction view [3]

scenario, there are four types of modeling that are run, namely S0 1-2, S0 1-3, S0 4-2, and S0 4-3, which are distinguished from the variable number of passengers, headway arrival of LRT, occupancy, and capital split factors. In the initial plan scenario, this will be analyzed using the measurement area and Pedestrian Travel Time approach (Fig. 2).

4.3.1 Measurement Area Analysis

The picture is the grid cells level of service in each simulation that has been done, where the blue indicates LOS A, blue-green indicates LOS B, green indicates LOS C, yellow indicates LOS D, orange indicates LOS E and the red is LOS F. From the results of the modeling, it can be concluded that some sidewalk locations get a poor level of service. This is caused by a large number of pedestrians that are not balanced with the condition of the sidewalk as supply. The level of service obtained refers to the value of the pedestrian flow and the modulus of space that has been previously treated. In this scenario, the average pedestrian generally experiences blockages in the sidewalk corner area, because modeled pedestrians prefer the fastest way to get to the destination so that the flow of pedestrians passing through the sidewalk network is very low. The worst conditions occur in simulation S0 4-2, where pedestrians lose space to walk to the destination according to the specified pedestrian route (Fig. 3).

4.3.2 Time Travel Analysis

Previously, a validation test had been carried out on the time needed by the pedestrian to reach his/her destination. For example, in Ped. Time Msr. Point 3, it only takes 420 s, but during simulation S0 1-2, it takes a longer time, which is 610 s. This

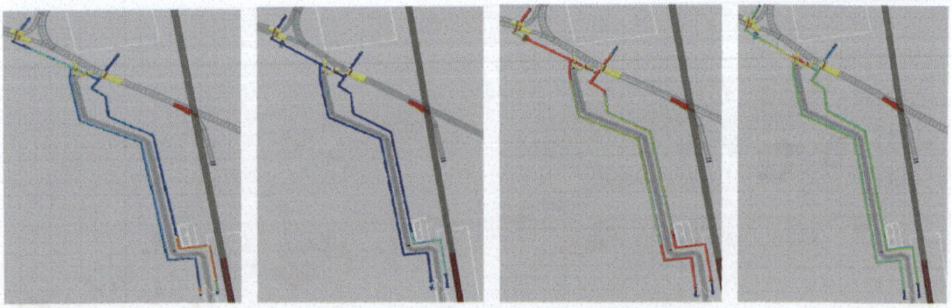

(a) (b) (c) (d)

Fig. 3 Grid cells results from PTV Vissim 11 in every simulation; **a** S0 1-2, **b** S0 1-3, **c** S0 4-2, and **d** S0 4-3

happens due to a slowdown caused by the obstruction of pedestrians on the sidewalk network.

4.4 Proposed Design Scenario (S1)

In the proposed scenario, some results of running an analysis of the initial scenario (S0) that has been modeled before are used as references to improve the condition of the level of service in each measurement area needed, such as the value of the effective width value (We). In addition, widening the corners of the sidewalk is held so as not to form a perpendicular angle, by widening the angle of the sidewalk. This is done so that the flow of pedestrians remains smooth when crossing the sidewalk network.

4.4.1 Measurement Area Analysis

If we see the results of the level of service in this proposed scenario, in general, the level of service at each measurement area increases. For example, in measurement area 3, which previously had a service level/LOS C, it changes to LOS A with a previous density of 0.42–0.19 p/m^2. This occurs due to several changes made, so that it can increase the level of service on the sidewalk network model. Figure 4 is the area grid cell scheme in each component of the measurement area.

4.4.2 Travel Time Analysis

The amount of travel time carried out by pedestrians has further diminished. For example, in Ped. Time Msr. Point 9, it previously needed around 734–662 s. This is

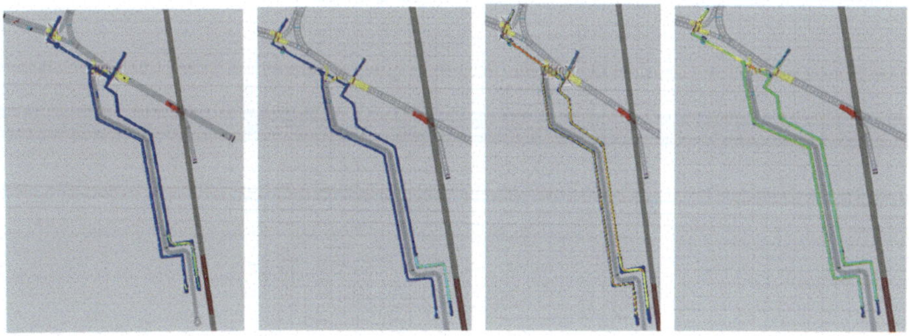
(a) (b) (c) (d)

Fig. 4 Grid cells results from PTV Vissim 11 in every simulation; **a** S1 1-2, **b** S1 1-3, **c** S1 4-2, and **d** S1 4-3

certainly due to changes in the layout and width of the sidewalk that is carried out to smoothen the flow of pedestrians on the sidewalk network.

4.4.3 Model Comparison

If the measurement area results are compared, you will see an increase in the desired level of service by changing the initial layout part of the plan, such as widening the sidewalk and widening the corners of the sidewalk meeting. This level of service is obtained from the results of the density level of space available as demand, namely LRT passenger volume which will go up and down at Cibubur Station, so that it can be concluded overall that the design of the modeled proposal can be a reference material for contractor developers to adjust the plan conditions with existing demand conditions. Broadly speaking, a summary of measurement area comparisons in each simulation are shown in Table 5.

Furthermore, if compared to the travel time on each route of direction and destination in the two modeled scenarios, it can be concluded that overall changes in the condition of the sidewalk carried out can accelerate pedestrian travel time. All figures in Table 6 are expressed in units of seconds (s) which state the travel time.

After obtaining the calculation results from the initial scenario, Fig. 5 is a graph of the relationship between the service level and the effective walkway width (We) required at each service level.

5 Conclusions

From the results of the research and analysis that have been carried out above, several points can be summarized as follows:

Table 5 Summary of model comparison between simulation S0 and S1

Measurement area	Scenario simulation											
	1-2			1-3			4-2			4-3		
	S0	S1	Conclusion	S0	S1	Conclusion	S0	S1	Conclusion	S0	S1	Conclusion
1	LOS C	LOS A	Better	LOS A	LOS A	Neutral	LOS F	LOS E	Better	LOS E	LOS C	Better
2	LOS C	LOS B	Better	LOS B	LOS A	Better	LOS F	LOS E	Better	LOS E	LOS D	Better
3	LOS C	LOS A	Better	LOS A	LOS A	Neutral	LOS F	LOS E	Better	LOS F	LOS B	Better
4	LOS B	LOS A	Better	LOS A	LOS A	Neutral	LOS F	LOS E	Better	LOS D	LOS B	Better
5	LOS F	LOS C	Better	LOS D	LOS A	Better	LOS E	LOS A	Better	LOS C	LOS A	Better
6	LOS B	LOS A	Better	LOS A	LOS A	Neutral	LOS F	LOS D	Better	LOS D	LOS A	Better
7	LOS B	LOS A	Better	LOS A	LOS A	Neutral	LOS E	LOS E	Better	LOS D	LOS C	Better
8	LOS F	LOS C	Better	LOS D	LOS A	Better	LOS F	LOS D	Better	LOS D	LOS A	Better
9	LOS E	LOS B	Better	LOS D	LOS A	Better	LOS A	LOS A	Neutral	LOS B	LOS A	Better

Table 6 Summary of travel time results between simulation S0 and S1

Measurement point	Scenario simulation															
	1-2				1-3				4-2				4-3			
	S0	S1	Diff (S1 – S0)	Conclusion	S0	S1	Diff (S1 – S0)	Conclusion	S0	S1	Diff (S1 – S0)	Conclusion	S0	S1	Diff (S1 – S0)	Conclusion
1	703	651	−52	Faster	664	642	−21	Faster	721	730	9	Slower	758	680	−78	Faster
2	784	696	−88	Faster	726	680	−46	Faster	n/a	786	n/a	n/a	858	726	−132	Faster
3	610	570	−39	Faster	584	563	−22	Faster	637	635	−2	Faster	661	590	−71	Faster
4	703	610	−92	Faster	654	603	−51	Faster	736	694	−42	Faster	768	638	−130	Faster
5	616	538	−78	Faster	572	533	−39	Faster	615	604	−10	Faster	637	558	−79	Faster
6	177	127	−50	Faster	148	122	−27	Faster	150	134	−15	Faster	159	125	−34	Faster
7	20	15	−5	Faster	15	13	−2	Faster	61	12	−48	Faster	13	12	0	Faster
8	26	23	−3	Faster	23	22	−2	Faster	60	21	−39	Faster	21	21	0	Faster
9	734	662	−71	Faster	699	652	−47	Faster	1613	726	−887	Faster	783	679	−104	Faster
10	693	574	−65	Faster	611	567	−43	Faster	772	629	−143	Faster	681	586	−95	Faster
11	722	655	−67	Faster	693	655	−37	Faster	1024	716	−308	Faster	800	664	−136	Faster
12	24	13	−11	Faster	14	12	−2	Faster	23	12	−11	Faster	12	11	0	Faster
13	792	699	−93	Faster	740	689	−51	Faster	1210	778	−432	Faster	854	714	−139	Faster
14	704	607	−97	Faster	649	602	−47	Faster	826	687	−139	Faster	771	623	−148	Faster
15	609	532	−77	Faster	561	513	−48	Faster	734	579	−155	Faster	626	541	−84	Faster
16	179	125	−54	Faster	146	121	−24	Faster	1520	128	−1391	Faster	150	122	−28	Faster
17	24	21	−3	Faster	21	20	−1	Faster	20	20	−1	Faster	20	19	−1	Faster

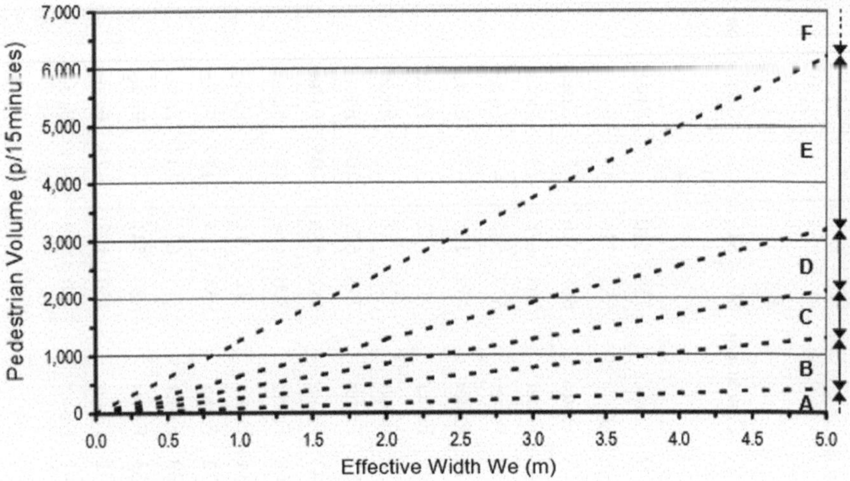

Fig. 5 Relationship between pedestrian volume and effective width walkways [4]

1. The initial design plan scenario (S0) of the sidewalk network from the main highway leading to the Cibubur LRT station is not feasible, and it can be seen from LOS E and LOS F in some measurement areas that have been simulated.
2. After several sidewalk design changes, such as widening the width of the sidewalk from the previous 2 to 3 m in several places or the required measurement area, the addition of the area in the corners of the sidewalk meeting seems to have very significant results in improving the quality of network services, which are given in Table 5, where almost all sidewalk networks have good service levels (A to C).
3. As a result of the sidewalk wide design changes that have been made in the proposed design scenario (S1), they also have an impact on the reduction in travel time needed by pedestrians on each pedestrian route. On average, it has a reduced travel time of up to 2 min. This is caused because the density of pedestrians has been further reduced, and the condition of the flow of passengers on the sidewalks to access the station has become more smooth. The overall results of the travel time comparison are summarized in Table 6.

Acknowledgements This article study is a part of research of International Indexed Publication for Universitas Indonesia Student's Final Project (PITTA) 2019.

References

1. Kittelson & Associates, I. *Transit Capacity and Quality of Service Manual.* (1999).
2. PUPR, K. (2013). *Perencanaan teknis fasilitas pejalan kaki.*

3. PTV Group. (2018). *PTV VISSIM user manual*. Karlsruhe: PTV Group.
4. Setiawan, R. (2005). Usulan Standar dan Evaluasi Tingkat Pelayanan Walkway di Universitas Kristen Petra. *DIMENSI (Journal of Architecture and Build Environment)* ISSN 2338-7858.
5. Highway Capacity Manual (HCM 2000).Transportation Research Board, Washington, D.C. 2000.

Service Quality Gap Analysis of Water Supply in Urban Areas: Water Supply Company, Bogor Regency

Armeinita Octavia and Chotib

Abstract High population growth in urban areas has led to higher activities which increase the needs of clean water. Efforts to provide water in urban areas come from the government known as water supply companies. The increasing demand will affect the quality of clean water itself, because of water resources capacity is limited and could create water service quality gap in the end. The purpose of this thesis is to identify the level of customer satisfaction with the service performance of PDAM Kota Wisata, to identify factors which influence service quality of water supply company and to analyse service quality gaps between water supply and the benefits received by the users (customers). This research uses a qualitative and quantitative approach, and data collection is done through area observation, agency visits and questionnaires, as well as literature reviews. The assessment of service quality is carried out through weighting based on the Likert scale, data processing and analysis methods which are carried out through the structuring equation modelling (SEM) method. The results showed some of the service quality aspects which are caused by several factors, start from the water resources capacity, the management of the water supply company and the operational standards and procedures in serving water to the users. The conclusions can be drawn by forming irreparable factors and give alternative solutions to reduce the gap of service quality between water supply and benefits to the users.

Keywords Water supply company · Service quality · Water resources · Gap analysis

A. Octavia · Chotib (✉)
Urban Studies, School of Strategic and Global Studies, Universitas Indonesia, Depok, Indonesia
e-mail: chotib@hotmail.com

S. S. Emamian et al. (eds.), *ICSDEMS 2019*,
https://doi.org/10.1007/978-981-15-3765-3_5

1 Introduction

The increasing of urban population requires the availability of clean water. SDG's number 6 challenge is to ensure the availability and management of sustainable water and sanitation for all, and then, the government's target is to achieve 100% clean water in 2019 which is not an easy thing to do without the help and coordination of various stakeholders, both in government and the community itself, including assistance from the private sector.

The management of Bogor Regency drinking water is managed by the regional water supply company (PDAM) Tirta Kahuripan which is designated as a water services BUMD (*Permendagri* 2/2007), and the management characteristics is non-rivalry and join consumption, but there are excludability (people will be charged for consuming water, the higher the usage, the higher the cost, and there is a market mechanism). And to achieve service for all, water management must be done properly, from water source, production and distribution system, including providing scales of services.

Permen PUPR 19/PRT/M/2016 has regulated cooperation in drinking water supply system (SPAM) implementation related to government support (DPP) on SPAM cooperation, procedures for granting government support and management of central/regional assets from government support. In case of funding support, it could be done by cooperation with the private sector (public-private partnership/KPBU).

PDAM *Tirta Kahuripan* classify customers as follows: I, IIA, IIB, IIIA, IIIB, IIIC, IVA, IVB, IVC, IVD, IVE and V. PDAM *Tirta Kahuripan* is a member of international organization (PERPAMSI), and its performance received rank tenth of healthy PDAM [1]. Based on *Permendagri* 23/2006, tariffs are set based on the principle of affordability and fairness, service quality, cost recovery, efficient use of water, transparency and accountability and protection of raw water/water resource, divided into four types of tariffs: (1) low tariff (tariff whose value is lower than the basic cost), (2) basic tariff (tariff with the same value or equivalent to the basic cost), (3) full tariff (higher value than the basic cost) and (4) agreement rates (tariffs which are based on agreements between PDAM and customers).

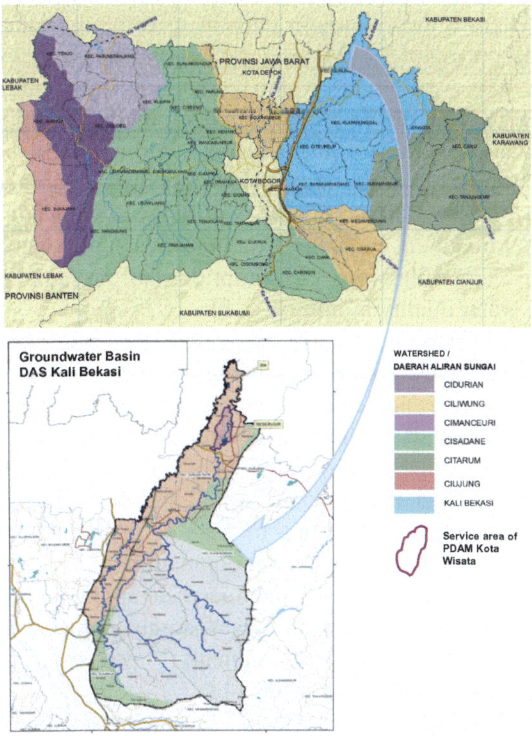

Two of PDAM *Tirta Kahuripan* service area located in separate watersheds of ground water basins (CAT) Bogor are Gunung Putri and Cileungsi (located in CAT Bekasi-Karawang) with the highest and third-highest population in Bogor Regency (7.9 and 6% of the total population in 40 sub-districts of Bogor Regency). The largest service area is managed by PDAM Kota Wisata Services Unit with 400 l/s capacity, and 120 l/s to serve Kota Wisata housing area (775 Ha), which is the largest and most integrated housing area with PDAM facilitation.

Although PDAM services for Kota Wisata has been integrated, and the group rates are included in IVA (luxury houses), there are still problems, both in water quality and service system that affect customer satisfaction. Some of the alleged problems that occur include the diminishing water discharge capacity received by users, frequent turbidity of water (especially during the dry season) and the absence of notification for repairing leaks, so it cannot be anticipated by the users. These problems occurred because of the increasing of PDAM users and declining infrastructure conditions.

The purpose of this study is to identify the level of customer satisfaction with the service performance of PDAM Kota Wisata, to identify factors which influence service quality of water supply company and to analyse service quality gaps between water supply and the benefits received by the users (customers). The result of this study will be useful as an input to PDAM Tirta Kahuripan in improving the quality of services, as well as for the local government in terms of regulating water supply policy for PDAM.

This study starts from the acknowledgement of drinking water supply system (SPAM) and the component of SPAM development (water resources, water management and water services distribution), which has been stated in the government regulation 122/2015, Permen PUPR 27/PRT/M/2016 stipulated the importance of water management, and it is inseparable from management principles in the form of utilization, protection and control, which are carried out in an integrated manner/multi-sector (Law 32/2009). This includes how to control the efficiency and sustainability of the drinking water treatment chain [2], existing water supply networks that highly stressed due to the increasing water consumption and the high quantity of water losses [3], also the water quality monitors [4, 5].

Urban water provision is clearly provided by water supply company (BUMD/PDAM), and it must create sustainable water resources management. But in the other hand, it still does not lead to a paradigm (mindset) of peoples' behaviour towards water itself and start to think that water is an unlimited resource. Water consumption factors are influenced by individual factors [6]. The Ministry of Public Works and Housing has set the standards for domestic water requirements that currently apply in urban areas according to the size of the population [7], which is >60–100 l/person/day with >3000–20,000 population.

The SPAM pipeline network development needs to apply the principles of water governance [8]: (*) Approach: open and transparent, inclusive and communicative, coherent and integrative, and fair and ethical; (*) performance and operations: accountable, efficient, responsive and sustainable.

One of the descriptive methods to describe customer satisfaction level (in this case is PDAM service quality) is service quality (ServQual). Quality of service can also be defined as everything that focuses on the business to meet the needs and desires of consumers which are accompanied by the accuracy in delivering it, so that compatibility is matched with consumer expectations [9]. Thus, service quality can be defined as an analysis of the expectation achievement level of services received by the customers/consumers.

Service quality can be identified in five dimensions: (1) reliability, (2) responsiveness, (3) assurance, (4) empathy and (5) tangibles [10–13]. The assessment of service quality will be differed in the order of dimensions according to the needs of the study. For the gap analysis, one of five gaps in ServQual will be used [14]: Gap 5: expected service versus perceived service.

2 Methods

This study used mix-methods (quantitative and qualitative approach) with cross-sectional type of survey. The quantitative approach has been done by questionnaire distribution and structured interview. Choice of sampling came from the population characteristics that had been facilitated by PDAM for providing water. The sample size had been measured by using 90% reliability, which is 100 respondents, using Slovin formula from 8.568 population, and the choice of sample is proportional to size. The study variables consist of a dependent variable (Y) and independent variables (X). The dependent variable is service quality of PDAM Kota Wisata, and the independent variables are five dimensions of ServQual: (1) responsiveness (2 aspects, 4 questions); (2) reliability (4 aspects, 9 questions); (3) empathy (3 aspects, 4 questions); (4) assurance (3 aspects, 8 questions); and (5) tangible (2 aspects, 7 questions). The measurement of each independent variable is based on certain indicators and using Likert scale from 1 to 4.

The qualitative approach has been done by field observations, in-depth interviews and documentation. The observation consists of passive participation observation and gathering documents from the field related to the study. In-depth interviews consist of semi-structured interviews with PDAM Kota Wisata and Gunung Putri Installation (IPA Gunung Putri), using recording and transcription, and it is equipped with documentation.

Data analysis has been done through univariate and multivariate data analysis and triangulation. The statistical data will be analysed using SEM method with analysis moment of structure (AMOS).

3 Results and Discussions

The study location is at the end of DAS Kali Bekasi, CAT Bekasi-Karawang. According to Bogor Regency spatial plan, the area is designated as a high intensity urban zone. The low slope rate is only around 0–15%, making the area a potential for development, especially for urban settlements. But according to ground water basins data, the study location is designated as prone zones of greater 40–60% zone drop, with 250–600 mg/l chloride content. PDAM Kota Wisata is one of the service units of PDAM *Tirta Kahuripan*, Branch Cileungsi, which has one installation named IPA Gunung Putri, with Cikeas River as the water resource (surface water). The capacity is 400 l/s. It served several settlements, and the biggest one is at *Perumahan* Kota Wisata, which has 8.568 active connections (SR/*Sambungan Rumah*). The number of employees working at PDAM Kota Wisata is 15 people including one Head of the Service Unit. The 14 staff consists of: customer relations (including five meter-readers), engineering, engineering administration and cashier/financial administration.

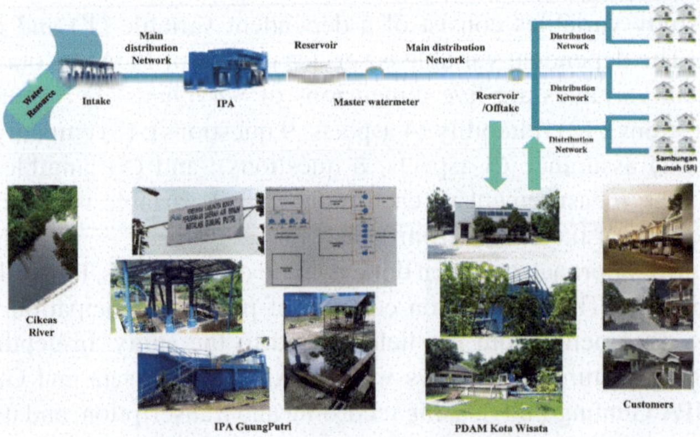

3.1 *Level of Customer Satisfaction with the Service Performance of PDAM Kota Wisata*

The most important thing in this study is the respondents' assessment of PDAM service performance. There is a questionnaire which has been given to the respondents to measure the service quality based on Likert scale 1–4 (lowest to highest). It will be compared with the level of importance (respondents' expectation) (Tables 1 and 2). To describe mean score of each item, the indicators and variables are using class intervals based on the equation below:

$$\frac{(\text{Highest answer score} - \text{Lowest answer score})}{\text{Sum of class/categories}}$$

3.1.1 Service Performance: Responsiveness

In carrying out payments, the clerk is quite alert and fast to serve residents who want to pay the PDAM bill at the payment counter. This is also in accordance with the

Table 1 Basic interpretation of indicator scores in research variables

No.	Score	Interpretation of performance quality and expectation/importance
1	1–1.75	Very low
2	>1.75–2.5	Low
3	>2.5–3.25	High
4	>3.25–4.0	Very high

Table 2 Accumulations of respondents' satisfaction

Item of *X*		Satisfaction level				Mean	Performance quality
		SP	P	TP	STP		
Responsiveness (*X*1)	F	44	477	234	41	2.66	High
	%	5.53	59.92	29.40	5.15		
Reliability (*X*2)	F	116	889	594	192	2.52	High
	%	6.48	49.64	33.17	10.72		
Empathy (*X*3)	F	44	454	251	47	2.62	High
	%	5.53	57.04	31.53	5.90		
Assurance (*X*4)	F	79	789	558	166	2.49	Low
	%	4.96	49.56	35.05	10.43		
Tangible (*X*5)	F	73	776	467	77	2.61	High
	%	5.24	55.71	33.52	5.53		

standard, operational and procedures (SOP) for administrative and financial services for PDAM customers.

On the other hand, the importance of item 1 (customer administration services) is low because all respondents who have PDAM facilities have never dealt with customer administration services, especially in registering as new customers, because it has been integrated with housing construction, so that at the time of the handover of the house, the PDAM is installed.

3.1.2 Service Performance: Reliability

Easiness of payment at the counters has the highest satisfaction among others, because there are other alternatives besides the payment counter in the PDAM office, such as paying through Government Bank ATMs (*Mandiri*, BNI, BRI, BTN), *Alfamart* and *Indomaret*, Online (*Tokopedia, Bukalapak*), etc. Although it generally shows high-quality performance, there are several important and vital items for customer service that are still relatively low, such as the aspect of water quality and quantity. However, the PDAM has implemented SOP for clean water fulfilment in maintaining the quality and stability of clean water, as well as maintaining the smooth running of water and the physical quality of water, including: *Permenkes* 492/Menkes/Per/IV/2010 concerning Requirements for Quality of Drinking Water, then the arrangement of water discharge through a distribution pump adjustment for peak and off-peak hours, and water quality monitoring by laboratory tests (once a month). With this low satisfaction for water quality and quantity, the result shows that about 20% of the respondents have switched or added other water resource (ground water).

On the other hand, the importance of this reliability dimension is related to easiness for payment system, accuracy of water meter-reading and the stability of PDAM water quality and quantity.

3.1.3 Service Performance: Empathy

The respondents felt well received when making complaints, and the location of the office is very accessible. However, the important item 'Sensitivity of officers in providing information regarding tariff changes' has the lowest quality performance, and a lot of respondents did not receive notification of the tariff changes, even though PDAM has the SOP of service to customers related to tariffs: socializing through online media, distributing letter of notification regarding tariff changes and posting information at the payment counter.

On the other hand, 'Sensitivity of officers in providing information regarding tariff changes' becomes the highest expectations of respondents. This information problem may have occurred because of the coordination and information delivery system from PDAM to the users are still not optimal, which is due to conventional style of delivering information (still using letter of notification/non-online/non-digital), and PDAM does not have all the RW contacts on each cluster in Kota Wisata.

3.1.4 Service Performance: Assurance

The skills of employees and engineering officers, based on the survey results, did not encounter significant problems, but there were problems related to the attitude and friendliness of the officers in resolving complaints, if we compare it with the SOP of recruiting PDAM employees already refers to *Perda Kabupaten Bogor* 7/2007 about the Organizations and Staffing of PDAM *Tirta Kahuripan*, and it requires the employees to have the education, skills and expertise needed, with a maximum age of 35 years, who has passed the selection, and for the service section, they have been educated with various types of training.

Item of assurance which has the lower quality performance is the information of water supply system aspect (condition of water resources, infrastructures, management and monitoring system). It is also in accordance with the lack of the SOP regarding the information of water supply system, unless there are users who request it. There is also an evidence that around 36% of the respondents did not even know where the water source of PDAM was coming from. Even though lots of them could answer the source was from the river, but 42% of them failed to mention which one was the correct river.

Table 3 Measurement model evaluation phase 2 (second-order outer model)

Latent variables	Manifest variables	Loading/weight	Rule of thumb	Note	Rank
Service quality (*X*)	Responsiveness (*X*1)	0.927	0.500	Valid	2
	Reliability (*X*2)	0.882	0.500	Valid	4
	Empathy (*X*3)	0.926	1.500	Valid	3
	Assurance (*X*4)	**0.941**	**2.500**	**Valid**	**1**
	Tangible (*X*5)	0.85	0.500	Valid	5

3.1.5 Service Performance: Tangible

Tangible items do not affect the level of satisfaction too much, because there are many alternatives on the payment system. However, for some respondents who still relied on travelling to the payment counter, several assessments of dissatisfaction were more related to the condition of the service office that was less feasible, and there is no SOP for service facilities at the PDAM, only adjusting to service requirements, at least having a payment counter and a complaint office.

Although there are many alternative payments that can be done, the respondents' expectations regarding PDAM service office facilities still need to be improved, because some respondents still choose to pay manually to the counters, especially those who do a lot of activities around their houses.

3.2 Factors Identification Which Influence Service Quality of Water Supply Company

To get the key factors and indicators that have major roles in service quality, the method used is confirmatory factor analysis (CFA), using AMOS software as the analysis tool which is used in one of the structural equation modelling (SEM) procedures. With this analysis, it can be identified which indicators can explain a construct/concept/variable.

As we can see at Table 3, the aspect on assurance dimension has the highest loading factor (0.94), with the best indicators forming assurance variables are the skills of employees and technical officers.

3.3 Service Quality Gap Analysis Between Water Supply and the Benefits Received by the Users (Customers)

This gap analysis will give the information about how much the level of customer satisfaction for service quality is. This level of customer satisfaction can be determined by means of the average value of expectations then disputed by the average value of performance, which can result in three categories: (1) Positive (+), category: Very Satisfied; (2) Zero (0), category: Satisfied; (3) Negative (−), category: Not Satisfied.

The highest quality service gaps are found in the quality and stability of clean water, quantity (debit) and physical quality of water (water clarity), sensitivity of officers in providing information on tariff changes and information regarding the condition of water sources, conditions of water supply infrastructure, conditions of management systems and supervision of water use including the easiness of obtaining information related to PDAM policies.

4 Conclusions

PDAM Kota Wisata service is still not able to provide maximum service quality for its customers. The main aspect for improving service quality, based on the statistical analysis result, is the skills of employees and technical/engineering officers. But, it is not limited only on skills at payment process and technical implementation in the field, it also needs to improve attitudes in serving customers, to avoid conflicts and misunderstandings when complaints occur. This means it needs to improve the human resources, especially who serve directly to the users (costumers).

There is still dissatisfaction especially in the aspect of quality and quantity. This became a concern because although the planning has been integrated with PDAM facility, and has been equipped with several service SOPs, and it has been charged with luxury houses tariff, the problem still occurred. Further action must be done especially regarding improving better quantity and quality water provision, because it will straightly be related to water sustainability in the area.

On the other hand, the provision and management of clean water for urban areas cannot only be handled by PDAM, and it needs coordination from various sectors, from the government to the user community, including the involvement of private sector, so it can achieve the principle of water governance, which is accountable, efficient, responsive and sustainable.

Acknowledgements This article is part of Grant of International Publications for Students Final Project of Universitas Indonesia: Urban Clean Water Fulfillment.

References

1. BPPSPAM. (2017). *Buku Kinerja PDAM 2017*. Jakarta: Kementerian Pekerjaan Umum dan Perumahan Rakyat.
2. Bei, E., Wu, X., Qiu, Y., Chen, C., & Zhang, X. (2019). *A tale of two water supplies in China; Finding practical solutions to urban and rural water supply problems*. Beijing: Accounts of Chemical Research.

3. Caceres, R. G., Galvis, M. T., & Fajardo, J. F. (2019). Towards an efficient and sustainable planning of the drinking water supply chain. *Journal of Cleaner Production*, 394–401.
4. Kosaka, K., Koike, Y., Miyabayashi, Y., Saito, K., Asami, M., Sasaki, M., et al. (2019). National survey of utilization of continuous water quality monitors in water supply systems in Japan. *Water Supply*, 1347–1353.
5. Soares, J., Leite, P., Teixeira, P., Lopes, N., & Silva, J. P. (2019). *Data warehouse for the monitoring and analysis of water supply and consumption*. Switzerland: Springer.
6. Hendriawan, F., & Cahyadi, A. (2016). *Analisis Faktor-Faktor Yang Mempengaruhi Konsumsi Air Bersih Pelanggan Kelompok Rumah Tangga Menengah (K3B) Di PDAM Tirta Kahuripan Kabupaten Bogor Wilayah Cabang Pelayanan XI Cibinong*. Kabupaten Bogor: STIE Dewantara.
7. Meutia, A. (2013). *Arahan Pengembangan Sistem Penyediaan Air Minum Kota Bogor*. Bogor: Institut Pertanian Bogor.
8. Batchelor, C. (2008). *Water governance literature assessment*. London: IIED.
9. Purnamawati, E. (2012). *Analisis Kualitas Layanan dengan Metode Servqual dan AHP di Dinas Kependudukan dan Pencatatan Sipil di Surabaya*. Surabaya: FTI—UPN Veteran Jatim.
10. Parasuraman, A. P., Berry, L., & Zeuthaml, V. A. (1988). SERVQUAL: A multiple-item scale for measuring consumer perceptions of service quality. *Journal of Retailing*, 12–40.
11. Nugraha, L., Yuniar, & Harsono, A. (2015). Usulan Peningkatan Kualitas Pelayanan Jasa Pendidikan Bahasa Inggris Menggunakan Metode Service Quality (Servqual) di LBPP Lia Martadinata Kota Bandung. *Jurnal Online Institut Teknologi Nasional, 3*(1), 61–72.
12. Triwibowo, S., Rukmi, H. S., & Harsono, A. (2014). Usulan Peningkatan Kualitas Pelayanan Pada Kawasan Wisata Kawah Putih Perum Perhutani Jawa Barat dan Banten dengan Menggunakan Metode Service Quality (SERVQUAL). *Jurnal Online Institut Teknologi Nasional, 2*(1), 13–23.
13. Affandi, H., Zaki, M., & Azmeri. (2017). Pengaruh Kualitas Pelayanan Terhadap Kepuasan Pelanggan Pada Perusahaan Daerah Air Minum (PDAM) Tirta Mon Pase Kabupaten Aceh Utara. *Jurnal Teknik Sipil Universitas Syiah Kuala, 6*(3), 297–308.
14. Maass, W. (2011). *Servie management—Service quality*. Saarbrucken: Chair in Economics—Information and Service Systems (ISS).

Review on Research Methods in Performance-Based Building Design of High-Rise Residential Property

Sulfiah Dwi Astarini and Christiono Utomo

Abstract The success of a construction project depends on a successful design. Performance-Based Building Design (PBBD) concept in high-rise residential property considers building performance for user convenience. The concept should create a more sustainable design. However, designers often find several obstacles that can reduce the level of design success in the form of models and design methods. These studies use various methods. The diversities are qualitative–quantitative paradigm, primary–secondary data sources, and method of analysis. This study aims to explore the research methods of 16 papers published in several reputable journals. The results were presented in a diagram of a map research method. The map divides four quadrants through the *x*-axis namely the source of data (primary–secondary) and the *y*-axis namely the research paradigm (qualitative–quantitative). This is useful to help choosing the research method that will be used in future research.

Keywords Research methods · PBBD · High-rise residential

1 Introduction

In the construction industry, the performance approach in buildings is developed as a result of the use of performance concepts. It has a definition as a product behavior related to use in the building. This concept is referred to as a performance-based building design [1]. Related to planning building performance, there is an approach to achieve sustainable building design, using a concept tool called Performance-Based Building Design (PBBD) [2]. The reasons of interest in performance-based building are to increase innovation by articulating the performance of the building and by offering the relevant procurement of construction practitioners. Furthermore,

S. D. Astarini (✉)
Department of Civil Engineering, Institut Teknologi Sepuluh Nopember, Surabaya, Indonesia
e-mail: sulfiahdwiastarini@gmail.com

C. Utomo
Institut Teknologi Sepuluh Nopember, Surabaya, Indonesia
e-mail: christiono@ce.its.ac.id

S. S. Emamian et al. (eds.), *ICSDEMS 2019*,
https://doi.org/10.1007/978-981-15-3765-3_6

PBBD also provides a policy of innovating in fulfilling more effective and efficient performance requirements [3].

Many researches discusses about the PBBD concept. The first is the functionality. They are accessibility, operation and maintenance building. The second is safety, containing structural safety and fire safety. The third is health and well-being. In the concept, there are indoor air quality, indoor climate, acoustics, and water quality. The last concept is sustainability. It containing energy efficiency, environmental impact, and durability.

The concept of PBBD is very interesting so that it's currently very widely used in construction projects. There are several research methods used, for example, research [4] uses literature studies to gather information about methods in assessing building performance in facilitating sustainable design. Researchers of [5] use a case study that critically reviewed performance-based design of building design solutions by interacting in 3D geometry and evaluating energy consumption for air conditioners. Researchers of [6] used a survey method user for occupants to determine indoor air quality on post-occupancy building performance and perceptions collected and processed using AHP [7].

Further research is needed regarding the method in research on the concept. For this reason, this paper will focus on mapping the research methods used and aims to analyze the research methods most widely used in previous studies on PBBD. The results of this study are expected to be a reference for the selection of methods for further research.

2 Conceptual Background

In concept of PBBD, the main thing is to determine the performance of the building design according to the user/client needs [8]. High-performance designed buildings reflect design excellence. Fully in accordance with the principles of PBBD, the proposed framework provides the designer with a flexible tool for the description of the structural performance, where the acceptable probability of exceedance can be chosen by a team of decision-makers (designers, building owner, tenants), depending on the peculiarities of the structure, and is seen as an alternative to the use of prescriptive approaches [9]. Performance-based design is a process of performance requirements defined and integrated into building design [1]. Performance-based design consists of explorations supported by strategies to generate and analyze alternatives that address challenges with explicit objectives [10].

3 Research Method

It consists of quantitative and qualitative methods with primary and secondary data sources. The research reviewed case studies, literature reviews, IT Model, and surveys. Case study research examines many of the features of some cases that are individuals, groups, organizations, movements, events, or geographic units [11]. Case study research has many advantages because it can classify the thinking of researchers, and connect abstract ideas into reality based on the observed case. This technique produces complex explanations in the form of narrative stories about certain people and events [11].

Literature review builds upon a collection of knowledge ideas that can be learned from and builds upon what others have done [11]. Literature reviews are also useful in obtaining multiple variables as attributes that can affect results [12]. Reading the articles has the purpose of introducing the topic and turning to specific research questions that will be the main focus of the research, then it establishes the importance of research questions, linking to previous studies and then outlining the theoretical framework and defining the main concepts [11].

IT Model in PBBD research can use Analytical Hierarchy Process (AHP) and Building Information Modeling (BIM). AHP assists in making decisions characterized by multiple interrelated and competing criteria. The main feature of AHP is the inherent capability of systematically dealing with a vast number of intangible and non-quantifiable attributes, as well as with tangible and objective factors [13].

BIM is one of the technologies used in the architecture, engineering, and construction that is able to simulate all information in a development project in a three-Dimensional model. BIM is a shared knowledge resource for information about a facility forming a reliable basis for decisions during its lifecycle [15]. BIM has not only a database saving and providing necessary information for projects, but also comprises many important functions for building performance analysis [15].

The survey technique is the most widely used technique because this technique can be used for exploratory, descriptive, and explanatory research that has several forms such as interviews, internet opinion polls, and various types of questionnaires [11].

4 Result and Discussion

It was found the articles are considered relevant to the topic of PBBD. However, previous research related to high-rise residential buildings is rarely discussed. The literature review is used to summarize the literature, identify gaps, and provide a framework for future research. In this study, the literature used is from journals of Advanced Engineering Informatics, Building Environment, Building Research and Information, Built Environment Project and Asset Management, Design Studies, Energy and Buildings, Fire Technology, International Journal of Sustainable Engineering, Journal Construction Engineering Management, Journal of Engineering Design, Renewable, and Sustainable Energy Reviews.

Currently, there are many studies that discuss PBBD. The study uses different methods depending on the purpose of the research, for this part of the results will be described methods and objectives used in previous studies. Researcher of [13] using a survey to building owners with the aim of obtaining information on fire loads in building safety simulations.

Research of [14] developed a literature review on computer simulations that were used in the design process. Study by [5, 15] used BIM modeling to facilitate designers in optimizing sustainable designs. The results show that compared to traditional designs, PBBD on design optimizations that use BIM can minimize Life Cycle Cost (LCC) and Life Cycle Carbon Emission (LCCE) [15]. Researchers of [7] used a survey aimed at evaluating building designs based on respondents' perceptions of indoor air quality. Researchers of [7] used AHP to obtain important weights of various environmental quality attributes in a room whose results can help facility managers manage buildings with limited budgets. Good air quality shows an increase in welfare and effectiveness (increased productivity) for occupants [6].

Sexton and Barrett [3] developed a literature review exploring PBBD. The research results state that the PBBD requires a strategic focus that integrates buildings, business performance, and innovation and is in line with a sustainable high-performance building population [3]. A case study was used by [16] to produce many alternative design solutions using parametric modeling in support of the exploration of geometric designs. Since client quality plays a major role in enhancing green design performance, clients should participate effectively during the design process [10]. Therefore, effective design team attributes and client's qualities may increase the performance of the design green building performance in order to enhance building performance and reduce building impact on the environment [10].

There are six research methods used, and consist of case studies, literature review combine survey combine IT model: AHP and Computational Fluid Dynamics (CFD), Literature review combine survey, survey combines IT model: AHP, literature review, case study combines IT model: BIM. Table 1 presents the implementation of research method in previous researches.

Sources of data consist of primary and secondary data collection [12]. Primary data are collected by the authors through observation, experiments, surveys, interviews, daily reports, and case studies. Secondary data are collected data by others and has been available from books, libraries, and the web [17]. The research methods most widely used by the previous researches is the literature review, survey, and case studies.

Table 1 Literature review map of previous research methods

No.	Method	Previous research
1	Case study	[5, 10]
2	Literature review, survey, IT model: AHP, CFD	[4, 13]
3	Literature review, survey	[6, 12, 17]
4	Survey, IT model: AHP	[7]
5	Literature review	[1–3, 8, 14, 16]
6	Case study, IT model (BIM)	[9, 15]

The advantage of using the literature review method is the availability of various references that can be used in research so that making it easier for researchers to find research of material. The weakness of using the literature review is that the information obtained is not easy to be accepted and digested by researcher, and some sources of literature may have changed due to old issues. For the survey method, the advantage is relatively inexpensive, it can be determining the number of samples so that to give significant statistical results and its area was broad to reach remote locations by using mail, email, or telephone. The weakness of the survey method is enabling the researcher to obtain unwanted respondents, and researchers can ensure that most samples can provide the response. The advantage of the case study method is that it can provide as-is reports regarding specific events and not obtained from other researchers, whereas the weakness is that it less contributes to the practical problems in overcoming a problem.

Qualitative analysis provides a set of hypotheses ready to be tested from the literature. In this analysis, the researcher captures all the details of social settings in very detailed descriptions and conveys an intimate feeling for the settings and inner lives of people. Whereas quantitative analysis uses a set of interrelated constructs (or variables) that form propositions, or hypotheses, which determine the relationship between variables (usually in big terms or directions) and are presented by statistics and other programs for analyzing data. The mapping of research method from previous research in PBBD is shown in Fig. 1.

Based on Fig. 1, the most widely used research method is the analysis of quantitative data with primary data sources. The data is collected based on surveys and interviews from parties influencing the construction industry. The tendency of researchers to use quantitative analysis with primary data sources also began to develop due to the knowledge to develop existing hypotheses based on previous researchers.

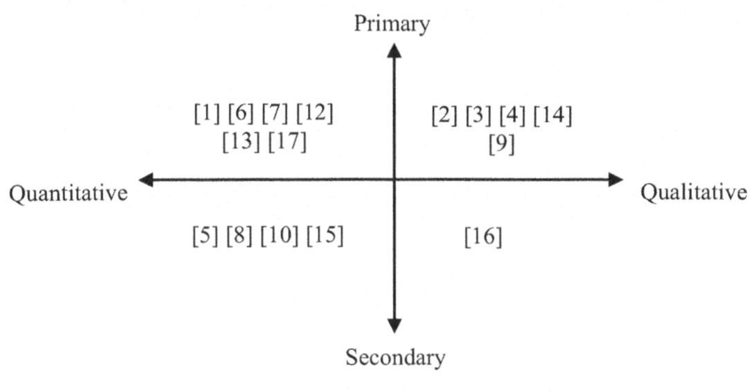

Fig. 1 Literature mapping of research method

5 Conclusion

From the description above, the research on the concept of design performance will continue to develop. Researches from the year 2008 to 2015 used qualitative methods with more exploring research on PBBD, then from the year 2011 to 2019, research methods on PBBD began to tend to use quantitative methods. This is due to researchers starting to develop hypotheses built from previous research.

It can be concluded that further research can consider qualitative primary methods, which can be carried out by questionnaire or interview, and analyze data by using statistical programs. After conducting a literature analysis, it also provides a better understanding of the approach methods that are mostly used to collect data. The results have also identified important questions in future research to find out the PBBD variable.

Acknowledgements The authors appreciate the recognition and awards in forms of research grants and fellowships from "Penelitian Dasar 2019 Kementerian Riset, Teknologi, dan Pendidikan Tinggi Republik Indonesia, based on contract number 845/PKS/ITS/2019".

References

1. Sayin, S., & Celebi, G. (2015). A review on performance based building design models. In *2nd International Sustainable Buildings Symposium* (962 pp.). Ankara, Turkiye: Gazi Universitesi.
2. Oxman, R. (2008). Digital architecture as a challenge for design pedagogy: Theory, knowledge, models and medium. *Design Studies, 29,* 99–120.
3. Sexton, M., & Barrett, P. (2005). Performance-based building and innovation: Balancing client and industry needs. *Building Research & Information, 33,* 142–148.
4. Iwaro, J. A., Mwasha, A., Williams, R. G., et al. (2014). The role of integrated performance model in sustainable envelope design and assessment. *International Journal of Sustainable Engineering, 8,* 294–316.
5. Rodrigues, E., Fernandes, M. S., Gomes, A., Gaspar, A. R., & Costa, J. J. (2019). Performance-based design of multi-story buildings for a sustainable urban environment: A case study. *Renewable and Sustainable Energy Reviews, 113.*
6. Thatcher, A., & Milner, K. (2016). Is a green building really better for building occupants? A longitudinal evaluation. *Building and Environment, 108,* 194–206.
7. Lai, H. K., & Yik, W. H. (2009). Perception of importance and performance of the indoor environmental quality of high-rise residential buildings. *Building and Environment, 44,* 352–360.
8. Becker, R. (2008). Fundamentals of performance-based building design. *Building Simulation, 1,* 356–371.
9. Bernardini, E., Spence, S. M. J., Kwon, D.-K., & Kareem, A. (2014). Performance-based design of high-rise buildings for occupant comfort. *Journal of Structural Engineering, 141.*
10. Clevenger, M. C., Haymaker, R. J., & Ehrich, A. (2013). Design exploration assessment methodology: Testing the guidance of design process. *Journal of Engineering Design, 24,* 165–184.
11. Neuman, L. W. (2014). *Social research methods qualitative and quantitative approaches* (7th ed.). Boston, MA: Pearson.

12. Korkmaz, S., Riley, D., & Horman, M. (2010). Piloting evaluation metrics for sustainable high-performance building project delivery. *Journal of Construction Engineering and Management, 136,* 877–885.
13. Zalok, E., & Hadjisophocleous, V. (2011). Assessment of the use of fire dynamics simulator in performance-based design. *Fire Technology, 47,* 1081–1100.
14. Ercan, B., & Elias Ozkan, T. (2015). Performance-based parametric design explorations: A method for generating appropriate building components. *Design Studies, 38,* 33–53.
15. Liu, S., Meng, X., & Tam, C. (2015). Building information modelling based building design optimization for sustainability. *Energy and Buildings, 105,* 130–153.
16. Turrin, M., Buelow, P., & Stouffs, R. (2011). Design explorations of performance driven geometry in architectural design using parametric modelling and genetic algorithms. *Advanced Engineering Informatics, 25,* 656–675.
17. Azari, R., & Kim, Y.-W. (2015). Integration evaluation framework for integrated design teams of green buildings: Development and validation. *Journal of Management in Engineering, 32.*

A Study of Non-destructive Test (NDT) Results on Old Concrete Specimens Under Loaded and Unloaded Conditions and Their Crushing Strength

Ajmal Paktiawal and Mehtab Alam

Abstract Concrete in the structures undergoes the environmental effect of changing weather during its life span and carries a certain minimum load. Some concrete structures suffer from damage due to weather effects of the environment at early age. Environmental effect with age on concrete specimens of different grades under loaded and unloaded conditions has been studied using; ultrasonic pulse velocity, rebound hammer, and crushing strength. A systematic planned experimental study (using ultrasonic pulse velocity and rebound hammer tests) has been conducted on unloaded and loaded concrete with 20% of expected crushing strength. The results of ultrasonic pulse velocity and rebound hammer obtained from concrete specimens of different strengths and ages have been correlated with their compressive strength. The experimental test results indicate that the pulse velocity response of concretes of different ages with regards to NDT and crushing strength show that lean mixes respond poorly under unloading and loading conditions. It can be also concluded that the average value of rebound hammer under loading and unloading conditions do not show a clear decreasing/increasing trend with increasing age. Based upon the experimental results it may be concluded that concretes of lower grades respond poorly with age due to environmental effect.

Keywords Non-destructive testing · Ultrasonic pulse velocity · Surface hardness · Compressive strength

1 Introduction

Concrete is by far the most widely used construction material in the world that looks like stone. The word "concrete" is derived from Latin concretus, meaning "to grow

A. Paktiawal (✉) · M. Alam
Department of Civil Engineering, Faculty of Engineering and Technology, Jamia Millia Islamia, New Delhi, India
e-mail: ajmal.paktiawal@gmail.com

M. Alam
e-mail: malam1@jmi.ac.in

S. S. Emamian et al. (eds.), *ICSDEMS 2019*,
https://doi.org/10.1007/978-981-15-3765-3_7

together." Concrete is a manmade building material that consists of aggregates (filler) embedded in a hard matrix (the cement or binder) that fills the gaps among the aggregate particles and glues them together [1]. Non-destructive testing (NDT) is a technique used by inspector engineers to detect defects in materials and structures, either during construction or in the life of service [2]. The non-destructive testing mechanism is practicable both for old and new structures. It is often necessary to check whether the structure is compatible with its design use [3]. According to ACI 228.2R Committee report, the non-destructive test method can be used to evaluate hardened concrete properties and conditions of the concrete in bridges, buildings, deep foundations, and dams. It can also be applied for quality control of new construction, quality assurance of repaired concrete structures, and also check older concrete for rehabilitation [4]. The ultrasonic pulse velocity test, as prescribed in non-destructive testing of concrete method of test IS: 13311 (Part 1)-1992, is a non-destructive test, can be used to check the quality without destructing the structure and assesses the homogeneity, presence of cracks, and voids in the concrete [5]. According to the ACI committee 228.1R report, a pulse sends a short-duration, high-voltage signal to a transducer, causing the transducer to vibrate at its resonant frequency. At the start of the electrical pulse, an electronic timer is switched on. The transducer vibration is transferred to the concrete through a viscous coupling fluid. The vibrational pulse travels through the member and is detected by a receiving transducer coupled to the opposite surface of the concrete. When the pulse is received, the electronic timer is turned off and the elapsed travel time is displayed. The direct path length between the transducer is divided by the travel time to obtain the pulse velocity through the concrete [6]. The dynamic Young's modulus of elasticity (E) of the concrete can be obtained from the pulse velocity and the dynamic Poisson's ratio. From the principle of elastic wave propagation, the pulse velocity is proportional to the square root of the elastic modulus [6]. The rebound hammer also known as Schmidt hammer or Swiss hammer test is also a non-destructive test that gives a convenient and rapid indication of the compressive strength of concrete. The test is sensitive and the results are affected by the location of the hammer's plunger. If the plunger is located over the hard-aggregate particles, it will show a higher rebound number, whereas, over a large air void or over a soft aggregate particle, a lower rebound number will occur. According to ASTM C805, ten readings for rebound hammer number must be taken for a test [6, 7]. The combined use of ultrasonic pulse velocity with rebound hammer was experimentally investigated for early age behavior of recycled concrete aggregate at a different water-cement ratio of 0.4, 0.5, 0.6, and 0.7. It was reported that compressive strength and UPV results for fresh aggregate concrete were higher as compared to recycled aggregate concrete. However, the rebound hammer results obtained for the recycled aggregate concrete were higher than normal concrete [8]. The old concrete cubes of different compressive strengths were collected from the university campus. After collection, all cubes were cured in water for 28 days. After this period, the cube specimens were tested for ultrasonic pulse velocity, rebound hammer, and compressive strength. The test result indicated that strength was obtained with an accuracy of 91% by rebound hammer and 78% by ultrasonic pulse velocity test [9]. This paper studied a comparative analysis between the rebound hammer and an ultrasonic pulse velocity test. Concrete of different mix proportions of 1:2:4 and 1:3:6 with a constant water-cement ratio of 0.45 were tested for the suggested ages of 7, 14, 21, and 28 days. From the test result, it was concluded that no significant variation existed between the rebound hammer as well as the ultrasonic pulse velocity test and inferred that both tests could be applied while assessing the quality and strength of concrete [10]. This paper evaluated compressive strength by using non-destructive techniques. In this research work, concrete with varying compressive strengths from 20 to 50 MPa was cast and tested for 7, 28, and 56 days. The result showed that more age resulted in higher ultrasonic pulse velocity as well as rebound hammer and also increased the UPV as well as rebound hammer by reducing the water-cement ratio [11, 12]. This paper investigated damage evaluation of reinforced concrete buildings after seismic events. The research work carried out by shaking table and generated several earthquakes. During the investigation, direct and indirect sonic methods, as well as partial and complete ultrasonic tomography tests were carried out. These NDT tests were applied before and after shaking the table test. Considerable correlation between the inter-storey drift ratio and ultrasonic pulse velocity reduction in orthogonal direction of each column was recognized. A feasible procedure for inspection of RCC building subjected to earthquake was formulated. From the test result, it was found that the high difference in ultrasonic measurement was identified at mid-height and bottom of the column after seismic event [13]. This research work was carried out to study non-destructive tests on fine and coarse recycled aggregate concrete. In this paper, the combination of both fine and coarse recycled aggregates was replaced by natural aggregate at varied percentages of 0, 8, 20, and 31%. In this investigation, the influence of recycled aggregate on the correlation between E-Modulus and compressive strength was studied by UPV and electrical conductivity test. From the test result, it was concluded that for the same value of NDT, the higher was the amount of recycled aggregate, the lower was E-Modulus. At an early age, the correlation between compressive strength and E-Modulus was affected by the amount of recycled aggregate [14]. This paper studied the estimation of self-compacting concrete strength using a non-destructive test. Concrete of different water-cement ratios ranged from 0.3 to 0.4 were provided. Silica fume type sika fume-HR was used to enhance the stability of concrete. For rebound hammer, the cube was fixed under the compressive testing machine and applied a constant pressure of 5 MPa. From the test result, it was concluded that ultrasonic pulse velocity and rebound number were decreased with an increase in the water-cement ratio. It was also inferred that combined usage of rebound hammer and ultrasonic pulse velocity improved prediction in terms of compressive strength [15]. This paper reported the effect of stress on the result of ultrasonic pulse velocity in concrete. The concrete cylinders of different composition were subjected to gradually increased and repeated compression loading. In this investigation, the UPV test was measured under the loaded and unloaded conditions with different frequencies. From the test result, it was concluded that at gradually loaded stress level, the UPV measurements slightly increased, whereas, at 70% of its ultimate strength, it remained constant, and beyond 70% of its ultimate strength, the UPV result rapidly decreased [16]. This paper investigated to study plain and polypropylene fiber reinforced concrete (PFRC) at elevated temperature to obtain compressive strength and also to assess the reliability of Schmidt's rebound hammer test at elevated temperature. The concrete cylinders of 100 mm diameter and 200 mm length were tested at ambient temperature as well as the elevated temperatures of 200, 400, 600, and 800 °C. From the test result, it was found that the deviation value of plain concrete at elevated temperature was higher as compared to polypropylene fiber reinforced concrete. It was also concluded that Schmidt's rebound hammer test result was in good relation with the destructive compressive test result at an elevated temperature of less than 600 °C. And beyond the 600 °C, the rebound hammer test was not suitable [17].

2 Significance of the Research

For health monitoring and assessing the strength of the old concrete, ND Testing is required to be done, at times distressed structures are also assessed using non-destructive techniques such as ultrasonic pulse velocity and rebound hammer. Keeping this in mind, virgin concrete specimens of 150 × 150 × 150 mm of various concrete grades and varying age lying in the Engineering Materials Labs were collected and grouped as shown in Table 1. These samples of concrete grades M25, M30, M40, M45, and M50 were cast in different years. A total number of 64 collected concrete specimens were segregated into twenty groups according to their age and grade. Each segregated group contained three or four specimens.

Table 1 Number of segregated specimens into different groups

Grade of concrete (MPa)	No. of specimens	Age (days)	Group No.	Grade of concrete (MPa)	No. of specimens	Age (days)	Group No.
50	3	272	G-1	40	3	165	G-11
	3	353	G-2		3	183	G-12
	3	387	G-3		3	406	G-13
	3	448	G-4		4	475	G-14
	3	503	G-5		4	833	G-15
45	3	154	G-6	30	3	256	G-16
	3	348	G-7		3	265	G-17
	3	356	G-8		3	135	G-18
	3	377	G-9	25	3	470	G-19
	4	503	G-10		4	880	G-20

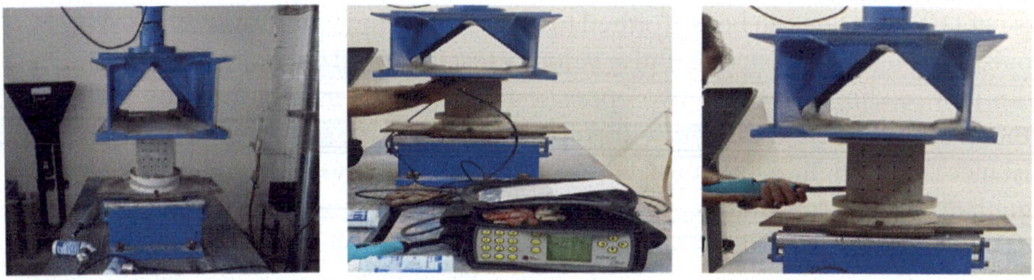

(a) Load applying (b) UPV reading (c) Rebound hammer reading

Fig. 1 Showing NDT test procedure on cube specimen of size 150 × 150 × 150 mm under loaded condition

3 Test Method and Procedure

In the present work, all specimens were tested for ultrasonic pulse velocity, rebound hammer, and compressive strength. The test procedure of concrete specimens was divided into two distinct conditions namely the loaded condition and unloaded condition. In unloaded condition, all specimens were tested for ultrasonic pulse velocity and rebound hammer, whereas, in the loaded condition, each specimen was placed under beam testing machine with capacity of 350 kN and the load applied from beam testing machine on the cube specimens without shock; and increased gradually at a constant rate of 14 $N/mm^2/min$ as specified by [18]. When the applied load from the beam testing machine reached 20% of its ultimate strength, then, beam testing machine was held and the specimen was first tested for ultrasonic pulse velocity and then for rebound hammer. After the UPV and the rebound hammer test, the same cube of size 150 × 150 × 150 mm was tested for compressive strength at a specified age. An automatic compressive testing machine with a capacity of 2000 kN was used. The load was applied without shock and gradually increased at a constant rate of 5.3 kN/s as given by [18]. These procedures are shown in Fig. 1.

4 Result and Discussion

In this part of the research, we discussed the results obtained from experimental tests, both for loaded and unloaded conditions. These results belong to ultrasonic pulse velocity, dynamic modulus of elasticity, rebound hammer, and compressive strength.

4.1 Ultrasonic Pulse Velocity

The ultrasonic pulse velocity test is the non-destructive technique of test which can monitor the quality of concrete without destructing the concrete member or structure.

As per [19], this test can be performed to assess the homogeneity, presence of cracks, voids, and quality of concrete. Pulse velocity test is carried out on four faces of cubes, with four readings on every two opposite faces. UPV was measured by taking the time in a microsecond that pulse traveled between two transducers across a known distance (Path length = 150 mm). Ultrasonic pulse velocity can be found as the known length divided by transit time. The natural frequency of transducers was 54 kHz. The UPV test was carried out under loaded and unloaded conditions. Under the loaded condition, all specimens were placed under a beam testing machine. A load of 20% of its ultimate strength was applied and then held the machine, and the pulse velocity was measured. Under the unloaded condition, the specimen was measured for UPV without any applied load. The results under loaded and unloaded conditions were presented in Tables 2 and 3 and plotted from Fig. 2a–d. By considering these figures, it can be concluded that UPV result for loaded condition for concrete of M50 grade which contained five different groups (G-1, G-2, G-3, G-4, and G-5) decreased by 2.4% as compared to unloaded condition, whereas, for concrete of grades M45, M40, M30, and M25, was decreased by 2.4, 3, 2, and 4.9%, respectively. The decrease in

Table 2 Unloaded condition results for ultrasonic pulse velocity test

Group No.	Age (Days)	Ultrasonic pulse velocity (km/s)				Average	Standard deviation	Coefficient of variation (%)
		*S*1	*S*2	*S*3	*S*4			
G-1	272	4.67	4.65	4.66	–	4.66	0.01	0.2
G-2	353	4.63	4.59	4.61	–	4.61	0.02	0.4
G-3	387	4.49	4.57	4.57	–	4.54	0.05	1.1
G-4	448	4.29	4.11	4.30	–	4.23	0.11	2.6
G-5	503	4.08	4.07	4.08	–	4.07	0.01	0.2
G-6	154	4.54	4.37	4.46	–	4.45	0.08	1.8
G-7	348	4.21	4.26	4.24	–	4.24	0.03	0.7
G-8	356	4.17	4.14	4.13	–	4.15	0.02	0.5
G-9	377	4.16	4.06	4.24	–	4.15	0.10	2.4
G-10	503	4.06	4.09	4.07	–	4.07	0.02	0.5
G-11	165	4.35	4.43	4.35	–	4.37	0.05	1.1
G-12	183	4.33	4.43	4.38	–	4.38	0.05	1.1
G-13	406	4.19	4.11	4.05	4.17	4.13	0.06	1.5
G-14	475	4.40	4.26	4.15	–	4.27	0.13	3.0
G-15	833	4.07	4.15	4.21	4.04	4.12	0.08	1.9
G-16	256	4.17	4.38	4.28	–	4.28	0.11	2.6
G-17	265	4.06	4.07	4.13	–	4.09	0.04	1.0
G-18	135	3.91	4.20	4.06	–	4.06	0.15	3.7
G-19	470	4.14	3.92	4.03	–	4.03	0.11	2.7
G-20	880	4.04	4.07	3.90	3.93	3.98	0.08	2.0

Table 3 Loaded condition results for ultrasonic pulse velocity test

Group No.	Age (Days)	Ultrasonic pulse velocity (km/s)				Average	Standard deviation	Coefficient of variation (%)
		*S*1	*S*2	*S*3	*S*4			
G-1	272	4.46	4.53	4.49	–	4.49	0.03	0.7
G-2	353	4.41	4.39	4.40	–	4.40	0.01	0.2
G-3	387	4.47	4.46	4.46	–	4.46	0.01	0.2
G-4	448	4.28	4.19	4.21	–	4.22	0.05	1.2
G-5	503	4.02	3.99	4.01	–	4.00	0.02	0.5
G-6	154	4.30	4.13	4.22	–	4.22	0.08	1.9
G-7	348	4.09	4.18	4.11	–	4.13	0.05	1.2
G-8	356	4.00	4.02	4.07	–	4.03	0.04	1.0
G-9	377	4.09	4.22	4.10		4.14	0.07	1.7
G-10	503	4.09	4.04	4.06	–	4.06	0.03	0.7
G-11	165	4.16	4.25	4.09	–	4.17	0.08	1.9
G-12	183	3.96	4.09	4.03	–	4.03	0.07	1.7
G-13	406	4.10	4.07	4.10	4.23	4.12	0.07	1.7
G-14	475	4.26	4.19	4.18	–	4.21	0.04	1.0
G-15	833	4.03	4.14	4.21	3.98	4.09	0.10	2.4
G-16	256	4.09	4.21	4.15	–	4.15	0.06	1.4
G-17	265	4.03	4.03	4.06	–	4.05	0.02	0.5
G-18	135	3.24	3.95	3.60	–	3.59	0.40	11.0
G-19	470	4.17	3.85	4.01	–	4.01	0.16	4.0
G-20	880	3.93	3.87	3.85	3.83	3.87	0.04	1.0

UPV under loaded conditions can be attributed to the propagation of internal micro-cracks caused by stresses in the concrete. In practical terms, structures under heavy loaded conditions, the pulse velocity test should not be used for the assessment of the quality of concrete.

4.2 *Dynamic Modulus of Elasticity*

Based on the obtained data for both loaded and unloaded conditions, now it is possible to find the dynamic Young's modulus of elasticity (E) of the concrete from the pulse velocity and the dynamic Poisson's ratio (μ) as specified by [19], using the following relationship:

$$E_d = \rho V^2(1 + \mu)(1 - 2\mu)/(1 - \mu)$$

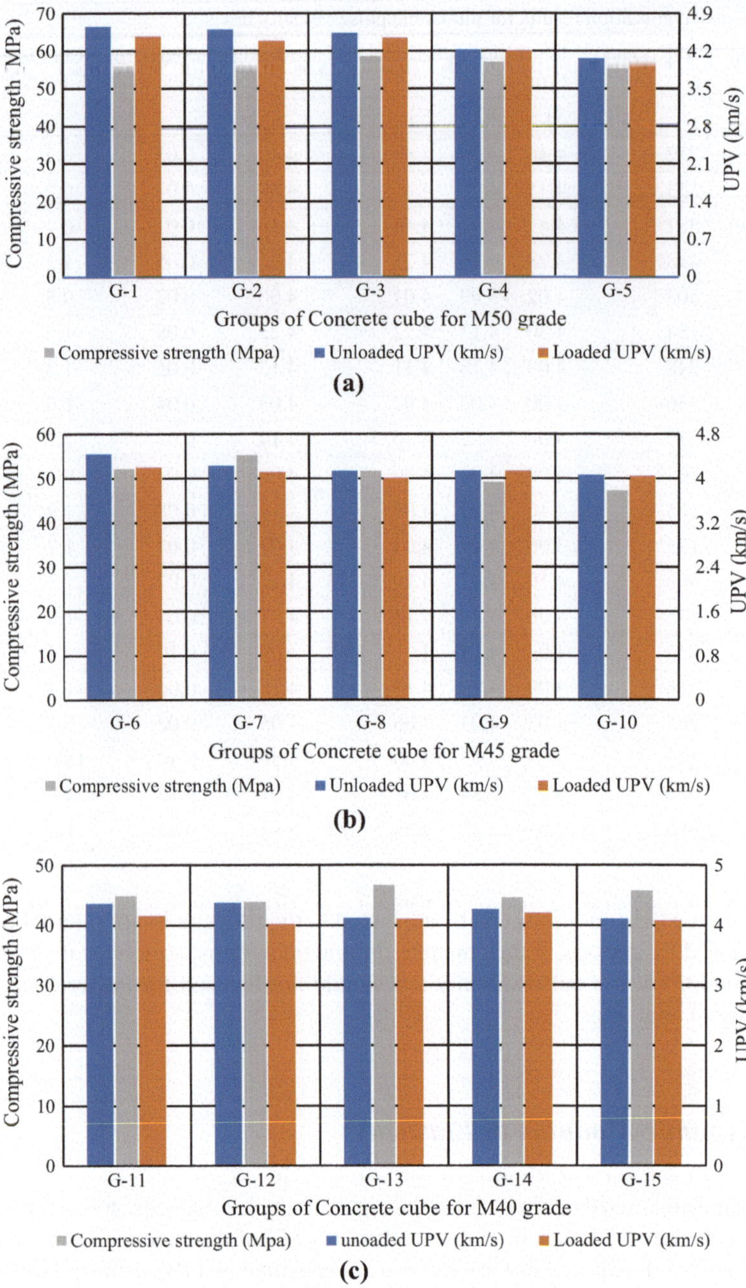

Fig. 2 **a–d** Relationship between compressive strength and ultrasonic pulse velocity

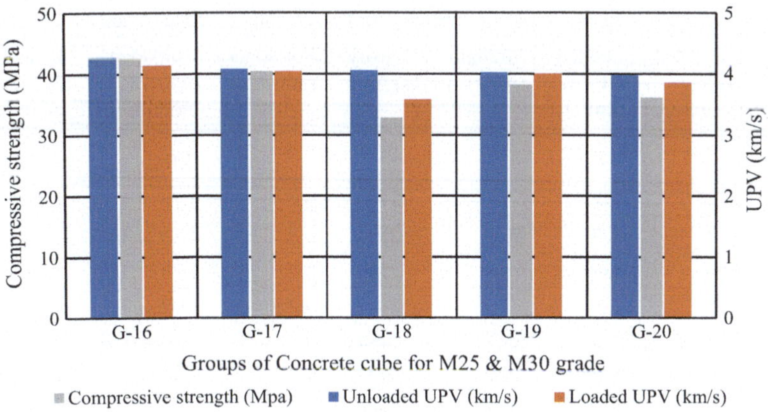

(d)

Fig. 2 (continued)

where

V is the pulse velocity in m/s
ρ is the density and
μ is the dynamic Poisson's ratio.

As per [19], Poisson's ratio varies from 0.2 to 0.35, which in this research work it is considered as 0.2. The dynamic modulus of elasticity of concrete is presented in Table 4.

4.3 *Rebound Hammer*

The rebound hammer test is a non-destructive test that gives a rapid indication of compressive strength of the concrete members. The aim of the rebound hammer is to assess the relationship between rebound hammer index and compressive strength as specified by [20]. In this investigation, rebound hammer tests were carried out on cube specimens of size 150 × 150 × 150 mm at different ages and compressive strengths. These specimens were first marked with a grid of 4 × 4@30 mm c/c having 16 points on each of the four vertical faces with marker. Ultrasonic pulse velocity and rebound hammer tests were conducted under unloading and loading conditions of 20% of expected crushing strength. In each condition of loading and unloading, alternate grid points were selected (Fig. 3) for UPV and rebound hammer. The results are presented in Tables 5 and 6 and plotted in Fig. 4a–d. From Fig. 4a–d, it can be concluded that rebound number value for a loaded condition for concrete of M50 grade had a negligible decrease by 1.2% as compared to unloaded condition, whereas, for concrete of grades M45, M40, M30, and M25 were decreased by 2,

Table 4 Dynamic Young's modulus of elasticity for loaded and unloaded conditions

Group No.	Unloaded UPV (km/s)	Loaded UPV (km/s)	Strength (MPa)	Density (kg/m^3)	Poisson's ratio	Unloaded dynamic modulus of elasticity (MPa)	Loaded dynamic modulus of elasticity (MPa)
G-1	4.66	4.49	55.8	2444	0.2	4.78×10^4	4.43×10^4
G-2	4.61	4.40	56.0	2534	0.2	4.85×10^4	4.41×10^4
G-3	4.54	4.46	58.5	2549	0.2	4.73×10^4	4.56×10^4
G-4	4.23	4.22	57.0	2497	0.2	4.02×10^4	4.00×10^4
G-5	4.07	4.00	55.3	2484	0.2	3.70×10^4	3.58×10^4
G-6	4.45	4.22	52.1	2423	0.2	4.32×10^4	3.88×10^4
G-7	4.24	4.13	55.2	2409	0.2	3.90×10^4	3.70×10^4
G-8	4.15	4.03	51.6	2438	0.2	3.78×10^4	3.56×10^4
G-9	4.15	4.15	49.1	2482	0.2	3.85×10^4	3.85×10^4
G-10	4.07	4.06	47.2	2529	0.2	3.77×10^4	3.75×10^4
G-11	4.37	4.17	44.8	24,440	0.2	4.20×10^4	3.82×10^4
G-12	4.38	4.03	43.8	2415	0.2	4.17×10^4	3.53×10^4
G-13	4.13	4.12	46.6	2395	0.2	3.68×10^4	3.66×10^4
G-14	4.27	4.21	44.5	2386	0.2	3.91×10^4	3.81×10^4
G-15	4.12	4.09	45.7	2463	0.2	3.76×10^4	3.71×10^4
G-16	4.28	4.15	42.4	2391	0.2	3.94×10^4	3.71×10^4
G-17	4.09	4.05	40.4	2469	0.2	3.72×10^4	3.64×10^4
G-18	4.06	3.59	32.8	2366	0.2	3.51×10^4	2.74×10^4
G-19	4.03	4.01	38.2	2399	0.2	3.51×10^4	3.47×10^4
G-20	3.98	3.87	36.1	2449	0.2	3.50×10^4	3.30×10^4

4, 1.4, and 2%, respectively. It can be also concluded that the probable accuracy of estimation of strength of concrete by rebound hammer in this investigation was + 11% and −4.4% as average and was within the acceptable region of clause 8.1 of [20].

5 Conclusions

In this research work, non-destructive test was carried out on old concrete cube under loaded and unloaded conditions, and based on the test results; the following conclusion can be derived.

- For all the groups, the UPV result under loaded condition is found to be lower as compared to the unloaded condition.

Fig. 3 Showing alternate grid points for loaded and unloaded conditions

- The average value of UPV under unloaded conditions is found decreasing for all groups with higher ages. It is due to the environmental effect of temperature variation on the concrete specimens.
- Average value of UPV under loaded condition does not show a clear decreasing/increasing trend with increasing age. It may be attributed to the development of new cracks along with random opening and closing of existing cracks due to environmental effects on concrete specimens.
- The average ultrasonic pulse velocity result under loaded condition for concrete grades of M50, M45, M40, and M30 is marginally lower by 2.45% as compared to unloaded condition, whereas, for concrete of grade M25 it is lower by 4.9% as compared to unloaded condition. It shows that the lower grade concrete is more susceptible to environmental damage than richer ones.
- The value of dynamic modulus of elasticity of concrete under loaded condition is found to be less than unloaded condition.
- For all the groups, the rebound hammer results under loaded conditions are found to be lower as compared to unloaded condition.
- The average of rebound hammer result under loaded condition for concrete grades of M50, M45, M40, M30, and M25 is found to be lower by about 2.1% as compared to unloaded condition.

Average values of rebound hammer under loading and unloading conditions do not show a clear decreasing/increasing trend with increasing age. Based on the experimental results, it may be concluded that concretes of lower grades respond poorly with age due to environmental effects.

Table 5 Unloaded condition results for rebound hammer test

Group No.	Age (Days)	Rebound hammer number				Average	Standard deviation	Coefficient of variation (%)	Compressive strength (MPa)
		*S*1	*S*2	*S*3	*S*4				
G-1	272	56.1	55.7	56.0	–	55.9	0.21	0.4	55.8
G-2	353	60.5	59.4	60.0	–	60.0	0.60	1.0	56.0
G-3	387	60.0	60.0	59.9	–	60.0	0.06	0.1	58.5
G-4	448	57.5	54.8	56.3	–	56.2	1.35	2.4	57.0
G-5	503	53.3	48.9	51.3	–	51.2	2.20	4.3	55.3
G-6	154	55.6	49.2	52.5	–	52.4	3.20	6.1	52.1
G-7	348	53.6	58.8	56.6	–	56.3	2.60	4.6	55.2
G-8	356	49.4	50.4	48.1	–	49.3	1.15	2.3	51.6
G-9	377	55.2	57.4	55.3	56.7	56.2	1.07	1.9	49.1
G-10	503	58.7	55.3	57.0	–	57.0	1.70	3.0	47.2
G-11	165	52.5	67.3	49.1	–	56.3	9.70	17.2	44.8
G-12	183	48.8	56.1	52.6	–	52.5	3.70	7.0	43.8
G-13	406	58.8	58.1	59.8	53.8	57.6	2.60	4.5	46.6
G-14	475	50.0	49.8	49.7	–	49.8	0.15	0.3	44.5
G-15	833	50.8	48.8	51.1	49.3	50.0	1.12	2.2	45.7
G-16	256	43.0	51.4	47.2	–	47.2	4.20	8.9	42.4
G-17	265	48.0	49.6	49.1	–	48.9	0.8	21.7	40.4
G-18	135	38.1	43.7	40.9	–	40.9	2.80	6.8	32.8
G-19	470	43.5	41.9	42.7	–	42.7	0.80	1.9	38.2
G-20	880	39.6	41.4	40.5	39.8	40.3	0.80	2.0	36.1

Table 6 Loaded condition results for rebound hammer test

Group No.	Age (Days)	Rebound hammer number				Average	Standard deviation	Coefficient of variation (%)	Compressive strength (MPa)
		*S*1	*S*2	*S*3	*S*4				
G-1	272	56.9	54.6	56.0	–	55.8	1.16	2.1	55.8
G-2	353	59.7	59.1	59.4	–	59.4	0.30	0.5	56.0
G-3	387	58.9	60.0	59.9	–	59.5	0.60	1.0	58.5
G-4	448	54.1	53.7	55.5	–	54.4	0.95	1.7	57.0
G-5	503	54.0	47.8	51.0	–	50.9	3.10	6.1	55.3
G-6	154	55.0	48.6	52.0	–	51.8	3.20	6.2	52.1
G-7	348	51.8	56.0	56.1	–	54.6	2.45	4.5	55.2
G-8	356	48.5	47.6	48.5	–	48.2	0.52	1.1	51.6
G-9	377	52.2	54.9	56.3	53.3	54.2	1.80	3.3	49.1
G-10	503	58.6	55.3	57.0	–	57.0	1.65	2.9	47.2
G-11	165	52.5	50.4	47.8	–	50.2	2.35	4.7	44.8
G-12	183	47.7	56.4	52.3	–	52.1	4.35	8.3	43.8
G-13	406	57.4	55.7	53.9	54.6	55.4	1.53	2.8	46.6
G-14	475	51.0	48.3	48.3	–	49.2	1.55	3.2	44.5
G-15	833	49.2	47.3	48.9	47.7	48.3	0.92	1.9	45.7
G-16	256	42.6	49.3	46.1	–	46.0	3.35	7.3	42.4
G-17	265	49.0	47.9	49.4	–	48.8	0.77	1.6	40.4
G-18	135	38.0	42.4	40.3	–	40.2	2.20	5.5	32.8
G-19	470	43.5	38.7	42.5	–	41.6	2.53	6.1	38.2
G-20	880	38.3	38.8	40.1	41.2	39.6	1.31	3.3	36.1

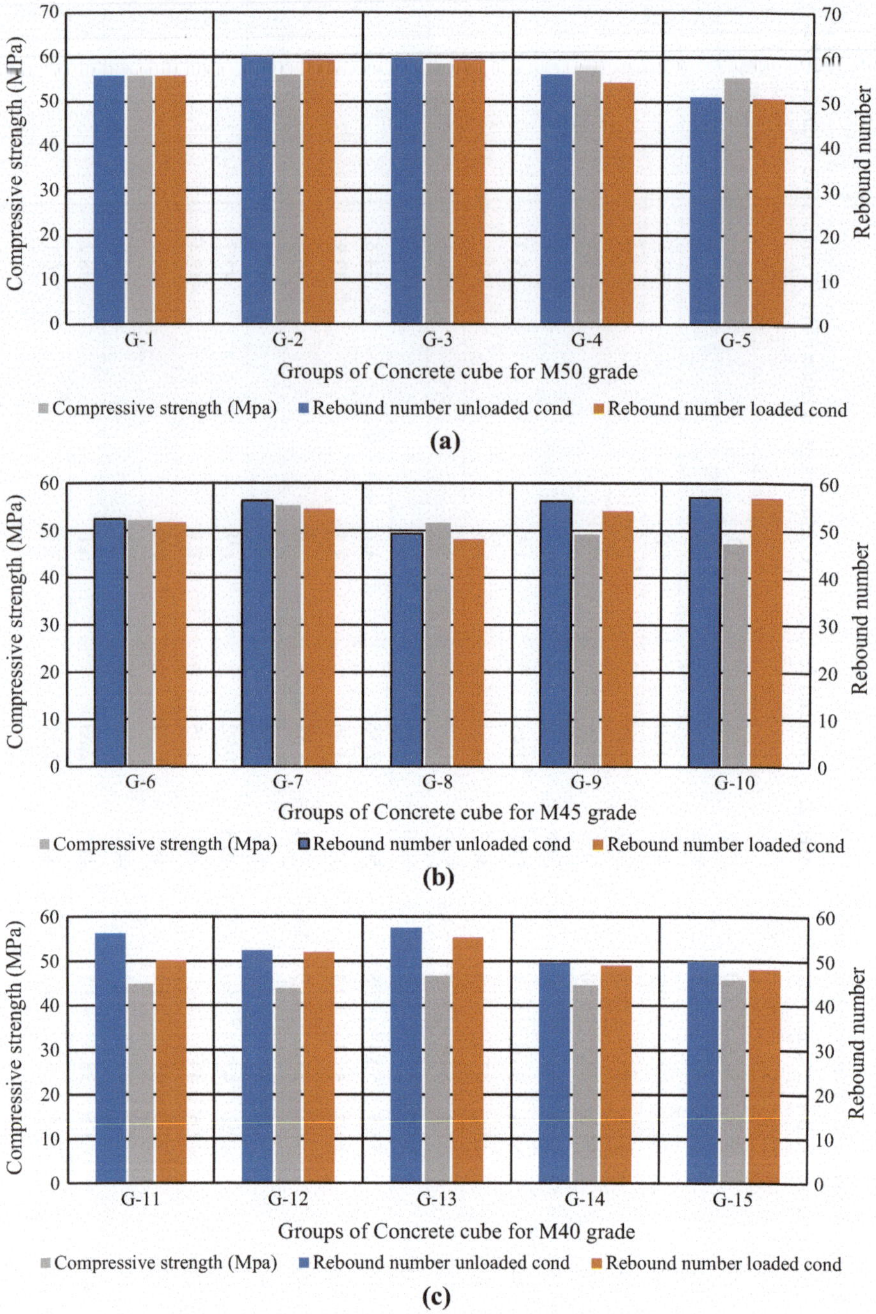

Fig. 4 **a–d** Relationship between rebound hammer number and compressive strength

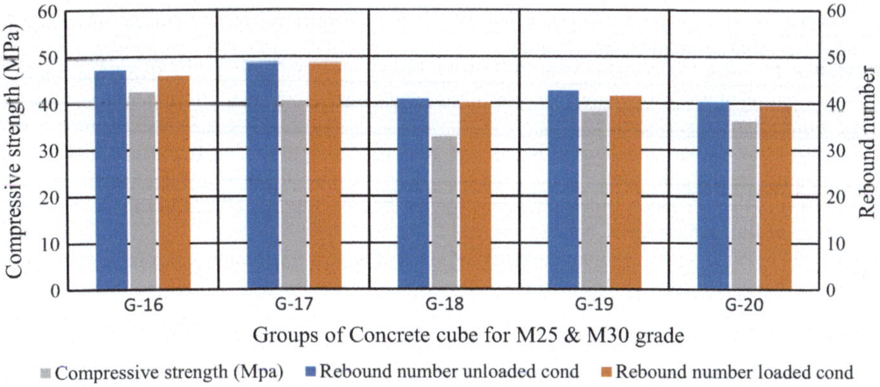

(d)

Fig. 4 (continued)

References

1. Li, Z. (2011). *Advance concrete technology* (pp. 1–506). Hoboken, New Jersey: Wiley & Sons, Inc.
2. Robert, A. S. (2015). An introduction to NDT common methods. *British Institute of Non Destructive Testing, 2,* 1–57.
3. International Atomic Energy Agency. (2002). *Guidebook on non-destructive testing of concrete structures* (Vol. 17, pp. 1–231). Vienna: International Atomic Energy Agency.
4. Committee, A. C. I. (2013). Report on non-destructive test methods for evaluation of concrete in structures. *ACI, 228*(2R), 1–98.
5. IS:13311 (Part 1)-1992, Indian Standard for method of test for ultrasonic pulse velocity test. Bureau of Indian Standards, New Delhi, 1992.
6. Committee, A. C. I. (2013). Report on in-place methods to estimate concrete strength. *ACI, 228*(1R), 1–44.
7. ASTM C805. (2018). Standard test method for rebound number of hardened concrete.
8. Al-Mufti, R. L., & Fried, A. N. (2012). The early age non-destructive testing of concrete made with recycled concrete aggregate. *Construction and Building Materials, 37,* 379–386.
9. Cavadar, A., & Bingol, S. (2014). Suggestion of new formulations for Schimdt hammer and UPV test method for concrete. In *Proceedings of the 3rd International Congress APMAS* (Vol. 125, pp. 402–404).
10. Agunwamba, J. C., & Adagba, T. (2012). A comparative analysis of the rebound hammer and ultrasonic pulse velocity in testing concrete. *Nigerian Journal of Technology, 31,* 31–39.
11. Rashid, K., & Waqas, R. (2017). Compressive strength evaluation by non-destructive techniques. *Journal of Building Engineering, 5,* 1–27.
12. Jain, A., Kathuria, A., Kumar, A., et al. (2013). Combine use of non-destructive test for assessment of strength of concrete in structure. *Procedia Engineering, 54,* 241–251.
13. Polimeno, M. R., Roselli, I., Luprano, V. A. M., et al. (2018). A non-destructive testing methodology for damage assessment of reinforced concrete building after seismic events. *Engineering Structures, 163,* 122–136.
14. Velay-Lizancos, M., Martinez-Lage, I., Azenha, M., et al. (2018). Compressive strength and non-destructive testing at early age. *Construction and Building Materials, 193,* 323–331.
15. Bzeni, D. K. H., & Ihsan, M. A. (2013). Estimating strength of SCC using non-destructive combined method. In *3rd International Conference on Sustainable Construction Materials and Technologies* (Vol. 8, pp. 1–8).

16. Popvics, S., & Povics, J. (1991). Effect of stress on ultrasonic pulse velocity in concrete. *Materials and Structures, 24*, 15–23.
17. Muller, P., Novak, J., Holan, J., et al. (2019). Destructive and non-destructive experimental investigation of polypropylene fiber reinforced concrete subjected to high temperature. *Journal of Building Engineering*, 1–15.
18. IS: 516-1959, Indian Standard for Method of Tests for Strength of Concrete. Bureau of Indian Standards, New Delhi, 1959.
19. IS: 13311 (Part 1)-1992, Indian Standard for Method of Test for Ultrasonic Pulse Velocity. Bureau of Indian Standards, New Delhi, 1992.
20. IS: 13311 (Part 2)-1992, Indian Standard for Method of Test for Rebound Hammer. Bureau of Indian Standards, New Delhi, 1992.

The Utilization of Polyethylene Terephthalate Plastic Bottle Waste for Asphalt Concrete Mix Modification

Amalia Firdaus Mawardi, Machsus Machsus, Rachmad Basuki, Mohamad Khoiri, Sukobar, and Ahmad Faqihul Muqoddam

Abstract The aim of this research is mainly to analyze the influence of adding polyethylene terephthalate (PET) plastic bottle waste to asphalt concrete-binder course mix. The polyethylene terephthalate (PET) plastic bottle waste was added to asphalt concrete-binder course mix using the wet processing method, in which the PET particles were directly mixed with the asphalt hot mix along with its aggregates. This whole research was conducted using asphalt with 60/70 penetration specification and 5.8% of asphalt content. While PET particles were added using various content ratios of 0, 1, 2, 3, 4, 5, 6, 7, and 8%. Based on the Marshall test, it is proven that the number of void in mix (VIM), viscosity modifying admixtures (VMA), and Marshall Quotient went up as the number of density, voids filled with asphalt (VFA), and Marshall flow went down.

Keywords Plastic waste · Polyethylene terephthalate · AC-BC · Marshall properties

1 Introduction

Waste management is an emerging challenge for Indonesia to solve. Recently, Indonesian waste production is estimated to reach 68 million tons by 2019, nearly 14% of the total waste is coming from plastic waste. Surabaya is estimated to produce 1500 tons of waste daily, in which 60% of the total waste was contributed from organic matters while the rest 40% were from inorganic matters like plastic [1]. Most plastic bottles were made from PET substance. This particular substance is difficult to recycle due to its complex chemical structure and demanding high technology machinery to break down into smaller particles to reuse. Therefore, it is commonly seen that plastic bottles are left unrecycled.

In order to keep a sustainable living, it is necessary to figure out a way to recycle PET plastic bottle waste. Utilizing PET waste as an asphalt concrete mix could be a

A. Firdaus Mawardi (✉) · M. Machsus · R. Basuki · M. Khoiri · Sukobar · A. F. Muqoddam
Department of Civil Infrastructure Engineering, Institut Teknologi Sepuluh Nopember Surabaya, Surabaya, Indonesia
e-mail: amaliafwirawan@gmail.com

S. S. Emamian et al. (eds.), *ICSDEMS 2019*,
https://doi.org/10.1007/978-981-15-3765-3_8

beneficial solution for both environment and society. Recently, the ministry of public work and housing republic Indonesia is planning to construct 258 km of brand new roads across Indonesian borders. This mega project would have an additional value for sustainability if the constructions were done using revolutionary method of adding PET into the asphalt concrete mix.

Generally, asphalt could be modified by adding polymer into its mixture. There are three polymers that are usually used to modify asphalt including elastomer thermo-plastic, plastomer, and reactive polymer. The addition of elastomer thermo-plastic could increase the elasticity of asphalt [2].

Asphalt concrete-binder course is an asphalt type that has very little influence on its surrounding weather and climate environment. Yet, it is required to have a certain thickness in order to stand the stretch and tension due to the traffic load. It is essential to take the asphalt stability into consideration during the mixing process [3, 4]. Based on the Indonesian national standardization code (SNI) 2010 annex 6.3.1 2c, it is required for asphalt concrete-binder course mix to have the maximum aggregate of 25.4 mm with a minimum thickness of 600 mm [5].

Ahmadinia et al. [6] did research on adding PET to asphalt concrete-binder course mix using a dry method in which the PET particles were added at the very end of the mixing process. It turned out that the optimum asphalt concrete was achieved at the level of 6%. During this particular process, the asphalt stability level and Marshall quotient were recorded to hit their highest level. The PET addition was proven to enhance the stiffness of the mix and resistance towards permanent deformation. The main intention of using the dry method is to preserve the PET with its original form which is also known for a semi-crystalline resin form so that PET has a minimum physical and property deformation.

Yasanthi et al. [7] re-summarized the previous research on waste utilization for asphalt construction. Yasanthi et al. [7] used sawdust, PET, and asphalt by-products for the research materials. Based on the loading and deformation test, it is concluded that the composition of the PET and sawdust at 2% from its total mass could be suggested as optimum support for roads with relatively low traffic.

Alfahdawi et al. [8] did an experiment on adding polyethylene terephthalate (PET), polypropylene (PP), and other plastic-related components to reinforced concrete, mortar, and substitution material for aggregation He concluded that, if a plastic-based material was added within the ratio of 0.1–0.5%, the mix property, and the durability could be improved by 14.28% while the compressive strength could be stimulated up to 34%.

2 Materials and Methods

The cylinder sample creation was done using asphalt concrete-binder course mix standard code. There were two steps of cylinder sample creation. First, the cylinder sample was prepared without adding the PET on the mixture to obtain the optimum asphalt content. Second, the cylinder sample by adding PET. However, the second step could be done right after the first step was finished in order to gain an initial insight into the optimum asphalt content. There are generally two types of asphalt mixing methods. There are wet and dry mixing processes. In general, the wet mixing process is better than the dry one in terms of preserving original properties. Yet, both methods should follow the minimum threshold of ductility [9].

2.1 Material and Tools

Various types and sizes of aggregates were used during this research. The coarse aggregate size ranges which were used during this research were (10–20 mm), (10–10 mm), and (5–10 mm) while the fine aggregates ere rock ash with the size of (0–5 mm). This research was also conducted using asphalt with 60/70 penetration specification (Fig. 1).

Fig. 1 Various sizes of aggregate used

The testing tools which involved during the testing process for this particular research included sieve shaker, Marshall tool, asphalt property test machinery, and other asphalt-related tools and the PET plastic bottle waste content ratio was varied as follows: 0, 1, 2, 3, 4, 5, 6, 7, and 8%. The PET plastic bottle waste was used as the additional material into the asphalt hot mix. Therefore, the PET plastic bottle waste had to be cut into small pieces using specialized crushed machinery in order to be blended with the asphalt hot mix.

2.2 *Material Testing*

Material test was done to examine and reassure the asphalt, aggregate, and additive substance qualities to check if they follow the standard code or not. The material testing process covered the test of physical aggregate property, physical asphalt property, and the additive. Tables 1 and 2 will portray the results of crushed aggregate and bitumen properties.

Table 1 Physical properties of aggregate

Properties	Standard code of testing	Value (%)
Abrasion test (%)	AASHTO T 96-87	28.38
Coating and stripping (%)	AASHTO T 182-82	95
Resistance to degradation (%)	AASHTO T 96-02	20

Table 2 Physical properties of bitumen

Properties	Standard code of testing	Unit	Value
Penetration	AASHTO T 49-68	1/10 mm	70.1
Ductility	AASHTO T 51-74	Cm	125.5
Ash and fire point	AASHTO T 48-74	C	270.5
Softening point	AASHTO T 53–74	C	52.5
Los and heating	AASHTO T 197-74	%	0.108
Specific gravity of asphalt	AASHTO T-100	gr/cc	1.030
Solubility asphalt	AASHTO T 44-70	%	99.59

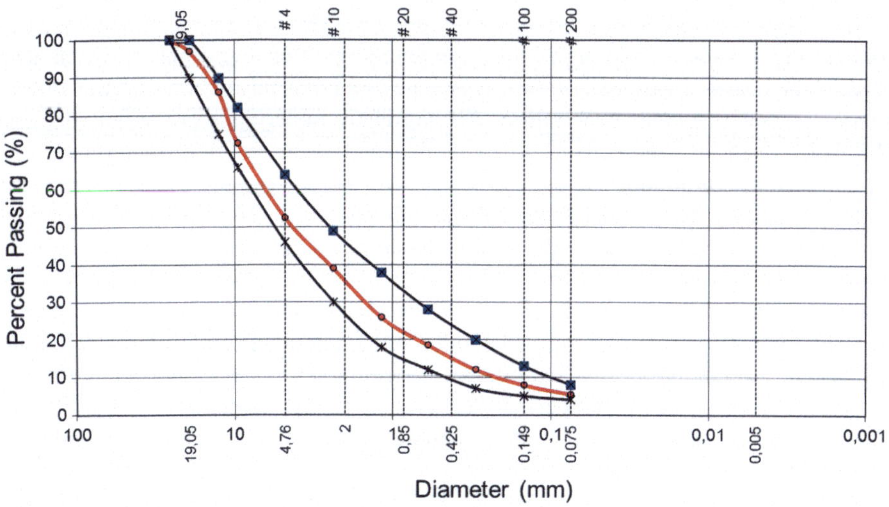

Fig. 2 The aggregate gradation of asphalt concrete-binder coarse

2.3 *Mix Design*

In order to obtain an ideal mixture design and optimum strength, it is necessary to compose a good mix design which follows either local or international specification and standard code. The aggregate mixing for cylinder samples must be done using a variety of material content in order to fulfill the Marshall parameter. Four fractions of combined aggregate were used during this research. Figure 2 will further explain which mixed gradation falls into the allowable limit.

2.4 *Marshall Test*

The aim of the Marshall test is to determine the durability, stability, and flow of asphalt mix based on AASHTO 245-90 standard code. The outcome of this test is a graph in which the results are plotted using the data of asphalt content and Marshall parameters which include asphalt density, void in mix (VIM), void filled with asphalt (VFA), void in mineral aggregate (VMA), stability, flow, and Marshall quotient.

3 Result and Discussions

All procedures and requirements done throughout this whole research were done using Indonesian national standardization (SNI) and the American Association of State Highway and Transportation Officials (AASHTO). The tests which were done during this research included the aggregate properties investigation which also covered the coarse, medium, and fine aggregate as well as physical properties investigation of 60/70 asphalt penetration. All aggregate materials were supplied from Mojokerto East Java area, while the asphalt penetration of 60/70 materials were supplied by PT. Bumindo.

There were a total of five cylindrical samples with each three varieties of asphalt contents that were prepared for the whole research. The exact proportion of asphalt content variety was obtained from a series of calculations and designs based on the Indonesia standard code. Each variety has interval difference of 0.5. Generally, all cylindrical samples weighted about 1200 g and composed of three different aggregates. All the aggregates and binders were mixed together under the temperature of 1550 °C, while the temperature for compaction process was set at 1450 °C. Compaction for asphalt concrete-binder course with relatively high traffic load will be further proceeded with total of 75 flows.

Right after figuring out the designed asphalt content, it is then processed into sampling preparation with interval difference of 0.5 for its asphalt content. Therefore, the cylindrical sample will have a variation of asphalt content 5,00; 5,50; 6,00; 6,50, and 7,00.

In order to obtain the optimum asphalt content, it is necessary to run the Marshall test first. Make sure that all materials were weighted under three different circumstances which include the dry weigh, saturated weight under the water, and saturated surface-dry. In order to obtain the saturated weight for aggregate, it is necessary to soak up the aggregate for 24 h then continue to go under the water bathing process for about half an hour under temperature of 600 °C. After that, run the series of Marshall tests to determine the stability and flow rate. Right after gaining stability and flow rate data, another Marshall parameter such as VMA, VFA, VIM, Density, and Marshall Quotient (MQ) could be further determined. Tables 3 and 4 will describe more about the Marshall parameter results.

Based on Table 4, it can be inferred that the asphalt optimum content was at 5.8% this percentage will further be used as the reference number to determine the asphalt optimum content in asphalt concrete-binder course mix with PET plastic bottle addition.

Table 3 Marshall parameter test result

Marshall parameter	Specification	Asphalt content				
		5.00	5.50	6.00	6.50	7.00
Stability	>800 kg	1436.9	1290.1	1227.8	1121.1	1267.9
Flow	(2–4) mm	3.11	2.79	2.71	2.79	3.18
VIM	(3–5) %	6.43	4.95	3.81	2.37	2.11
VMA	>14%	14.49	14.27	14.37	14.22	15.11
VFA	>65%	55.59	65.29	73.48	83.32	86.01
MQ	>250 kg/mm	461.53	462.96	453.07	401.81	398.28

Table 4 Determining optimum asphalt content

Marshall Parameter	Specification	Range	Optimum Asphalt Content (OAC %)
STABILITY	> 800 kg	5.0 - 7.0	
FLOW	(2-4) mm	5.0 - 7.0	
VIM	(3-5) %	5.5 - 6.0	
VMA	> 14%	5.0 - 7.0	
VFA	> 65%	5.5 - 7.0	
MQ	>250 kg/mm	5.0 – 7.0	
	Designed Asphalt Content (%)		**5.0** **5.5** **6.0** **6.5** **7.0**
	Optimum Asphalt Content (%)		**5.8**

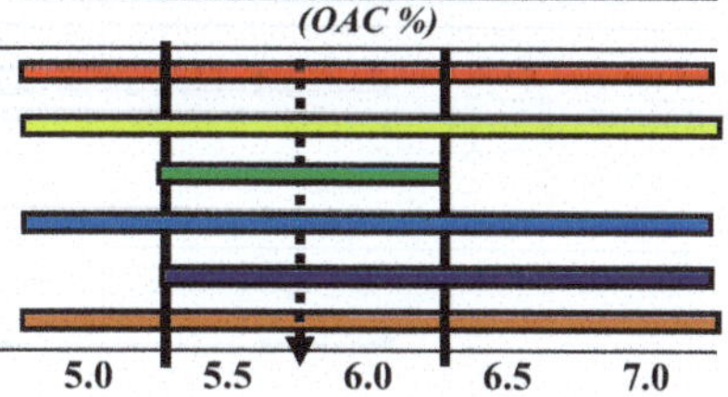

Aside from that, more cylindrical samples were then prepared with total of nine varieties based on the different PET contents. The varieties of PET contents were prepared at the percentage of 0, 1, 2, 3, 4, 5, 6, 7, and 8% from the asphalt total weight. This mixing process was done using wet process in which PET particles were added along with other aggregates and binders.

Figure 3 depicted that the density was inversely proportional with the plastic content. The more plastic content was added, the less density it had. There was a slight decline in the density value up to 5.0% of plastic content then it went back up to 7%. The highest density was reached at 2.42 when the plastic content showed its percentage at 1, 2, 7, and 8%. While the lowest density reached at 2.39 when the plastic content showed its percentage at 5%.

Figure 4 showed that the more PET plastic was added to the mixture, the more VIM it had, yet there was a slight decline at 6–8%. The highest VIM was hit at 6.26% when the plastic content showed at 5% rate. Based on this data, it could be inferred that PET plastic addition to the mixture could increase the number of voids in the mix.

Figure 5 describes that the more PET plastic was added, the more voids in mineral

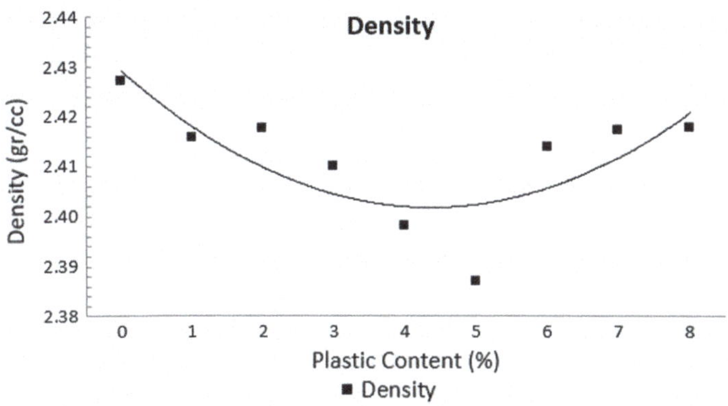

Fig. 3 Binder course density after PET addition

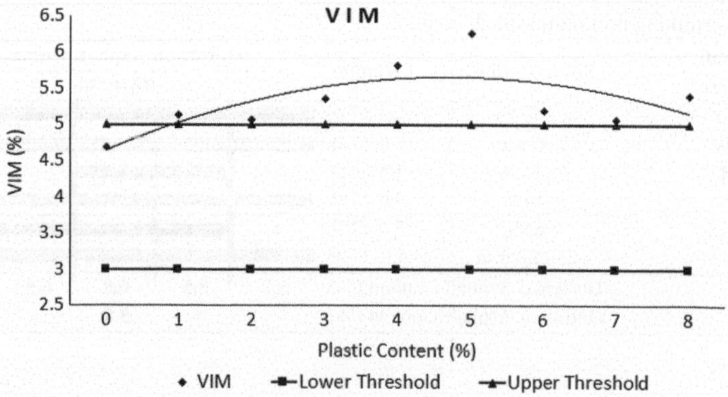

Fig. 4 Void in mix in binder course mixture after PET addition

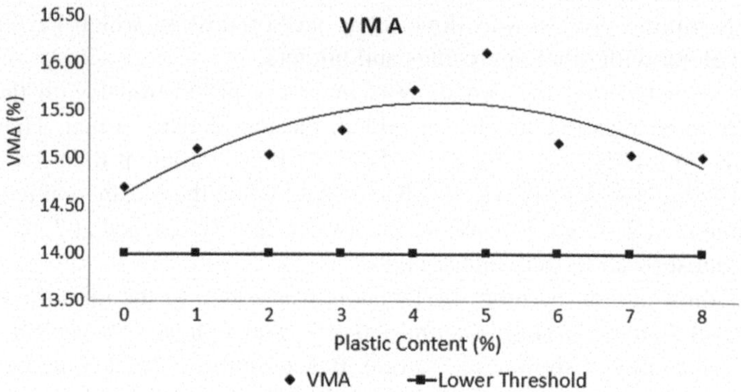

Fig. 5 VMA in binder course mixture after PET addition

aggregate were found. Yet, there was a slight decline at 6–8%. The highest VMA was at 16.11% when the plastic content was at 5%. While the lowest VMA hit at 15.02% when the plastic content was at 8%. Based on this data, it could be inferred that PET plastic addition to the mixture could increase the number of voids in mineral aggregate.

Figure 6 portrayed that the more PET plastic was added, the less voids in asphalt were found. Yet, there was a slight increase at 6–8%. The highest VFA was at 66.45% when the plastic content was at 8%. While the lowest VMA hit at 61.16% when the plastic content was at 5%. Based on this data, it could be inferred that PET plastic addition to the mixture could decrease the number of voids filled in asphalt.

Figure 7 showed that the more PET plastic was added, the more asphalt stability was found. Yet, there was a slight decrease at 4–8%. The highest stability was at 2286.6 kg when the plastic content was at 3%. The optimum stability was achieved at 2286.6 kg with plastic addition to its mixture. In contrast, the optimum stability

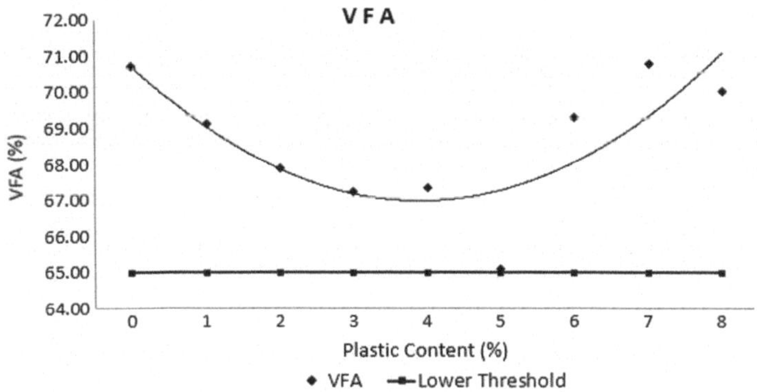

Fig. 6 VFA in binder course mixture after PET addition

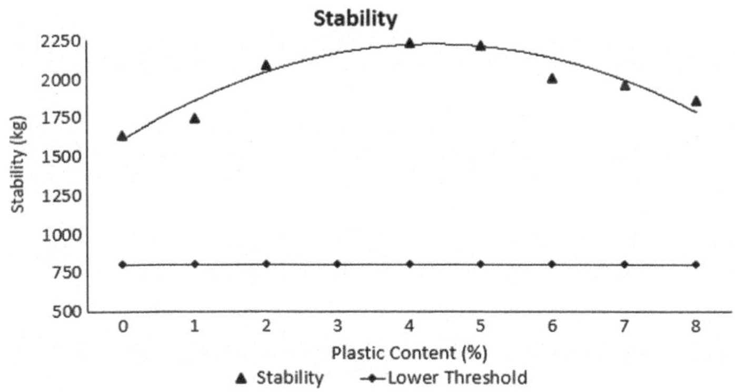

Fig. 7 Binder course stability after PET addition

was achieved at 1632.7 kg without plastic addition into the mixture. Therefore, the addition of PET plastic particles contributed to 40% increase in mixture stability.

Figure 8 depicted that binder course mixture which has PET addition, tends to have a small flow, yet its value went back up at the rate of 5–8%. The highest flow was at 2.52 mm when the plastic content was at 8%. While the lowest flow hit at 2.02 mm when the plastic content was at 4%. In contrast, the binder course mixture which has no PET plastic addition had the highest flow at 2.37 mm. This actually implies that PET plastic addition could boost up the flow rate by 6%.

Figure 9 presented that the more PET plastic was added, the bigger Marshall quotient value was recorded. Yet, there was a slight increase at 5–8%. The highest Marshall quotient value was recorded at 1103.35 kg/mm. Based on this data, it could be inferred that PET plastic addition to the mixture could increase the Marshall quotient value by 60%.

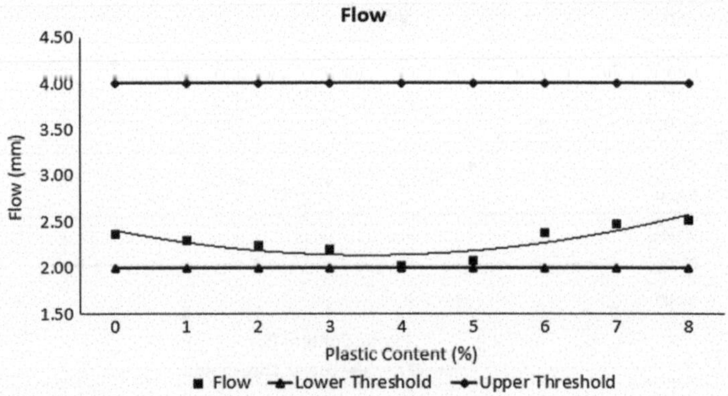

Fig. 8 Binder course flow after PET addition

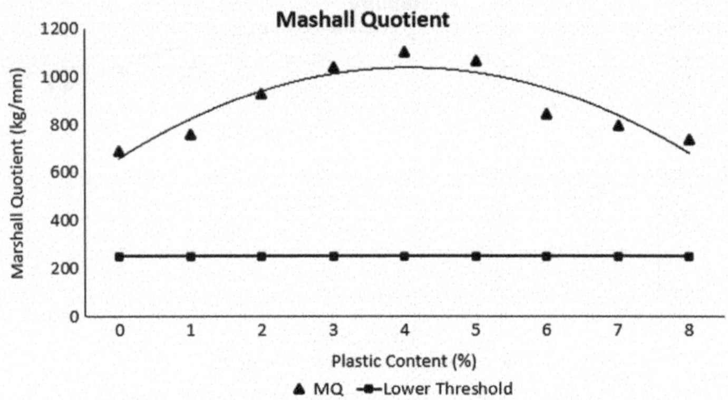

Fig. 9 Binder course's Marshall quotient after PET addition

4 Conclusions

According to the research experiments, it is inferred that the addition of PET plastic particles with the proportions of 1, 2, 3, 4, 5, 6, 7, and 8% into the asphalt concrete-binder course mixture could enhance the Marshall properties which include the stability, void in mix (VIM), void in mineral aggregate (VMA), and Marshall quotient itself. On the other hand, the PET plastic particle addition could also diminish the density, void filled in asphalt (VFA), and Marshall flow.

Acknowledgements We would like to express our gratitude to the reviewers for the humble constructive remarks. We also would like to thank the Institut Teknologi Sepuluh Nopember (ITS) Surabaya for the financial support provided through Research Grant.

References

1. Statistic of Surabaya Municipality. (2017). *Kota Surabaya dalam Angka*. Surabaya: Statistic of Surabaya Munacipality.
2. Modarres, A., & Hamedi, H. (2014). Effect of waste plastic bottles on the stiffness and fatigue properties of modified asphalt mixes. *Materials and Design, 61,* 8–15.
3. Sukirman, S. (2012). *Beton aspal campuran panas*. Jakarta: Granit.
4. Basuki, R., & Machsus, M. (2007). Penambahan Gilsonite Resin Pada Aspal Prima 55 untuk Meningkatkan Kualitas Perkerasan Hot Mix. *Journal Apl. Teknik Sipil, 3*(1), 16.
5. Marga, D. B. (2010). *Spesifikasi Umum Bidang Jalan dan Jembatan Revisi 3*. Indonesia: Kementrian Pekerjaan Umum.
6. Ahmadinia, E., Zargar, M., Karim, M. R., Abdelaziz, M., & Shafigh, P. (2011). Using waste plastic bottles as additive for stone mastic asphalt. *Materials and Design, 32*(10), 4844–4849.
7. Yasanthi, R. G. N., Rengarasu, T. M., & Bandara, W. M. K. R. T. W. (2016). Variation of Marshall properties of hot mix asphalt concrete with waste materials, pp. 75–81.
8. Alfahdawi, I. H., Osman, S. A., Hamid, R., & Al-Hadithi, A. I. (2016). Utilizing waste plastic polypropylene and polyethylene terephthalate as alternative aggregates to produce lightweight concrete: A review. *Journal of Engineering Science and Technology, 11*(8), 1165–1173.
9. Suroso, T. W. (2008). Pengaruh Penambahan Plastik LDPE, no. 3, pp. 208–222.

Analysis of Land-Carrying Capacity and Population Capacity Around Manado–Bitung Toll Road, North Sulawesi Province

Chairitha Rahmasani, Hafid Setiadi, and Komara Djaja

Abstract The Manado–Bitung Toll Road is one of the national strategic projects currently under construction and planned to start operating at the end of 2019. The existence of this Toll Road is expected to increase accessibility between two national activity centers and spur regional development. A consequence that can arise from the development of toll road infrastructure is the massive development of the surrounding areas, especially those that have direct access to the toll gate. On the other hand, regional development must not exceed the capacity to maintain environmental and social sustainability from uncontrolled development problems. Analysis of the carrying capacity and carrying capacity of the region is one of the efforts that need to be done. Based on the analysis results, 51% of the study area has a high level of development capability. Of the 21,852 ha. of the research area, 8357.18 ha. can be developed into settlements and can accommodate 1,671,435 people.

Keywords Carrying capacity · Population capacity · Land development capability · Regional development · Manado-Bitung toll road

1 Introduction

Infrastructure is considered as the main tool in the development of an area. Macroeconomically, the availability of infrastructure services affects the marginal productivity of private capital, whereas, in the context of the microeconomics, the availability of infrastructure services affects the reduction of production costs [1]. Infrastructure has

C. Rahmasani (✉) · K. Djaja
Urban Development Studies, School of Strategic and Global Studies, Universitas Indonesia, North Sulawesi, Indonesia
e-mail: chairitha.rahmasani@gmail.com

K. Djaja
e-mail: djaja_komara@yahoo.com

H. Setiadi
Department of Geography, Faculty of Mathematics and Natural Science, Universitas Indonesia, North Sulawesi, Indonesia
e-mail: hafid.setiadi@ui.ac.id

S. S. Emamian et al. (eds.), *ICSDEMS 2019*,
https://doi.org/10.1007/978-981-15-3765-3_9

an important influence on improving the quality of life and human welfare, among others in increasing the value of consumption, increasing labor productivity, and access to employment [2].

Hurst, [3] argues that accessibility is a measure of ease (time, cost, or effort) in moving between places or regions in a system. As for Black, [4], he is for the view that accessibility is a concept that combines a geographic land use management system with a transportation network system that connects it. The indicator in accessibility is a numerical value, which indicates the level of ease of getting access to goods and services [5].

Dennis, [6] states that there are at least three things that will be obtained if accessibility is improved, including (a) time savings, (b) reduction of transportation effort, greater freight capacity with longer distances, and (c) efficiency of movement and savings in transportation costs.

The Manado–Bitung Toll Road Development is one of the strategic projects that is expected to improve accessibility between two national activity centers, namely the City of Manado and the City of Bitung and spur development of the surrounding areas. The consequence that can arise from the development of toll road infrastructure is the massive development of the region, especially those that have direct access to the toll gate, so that changes in land use and land use change occur, even changes that are not in accordance with the regional spatial plan. Even though it is still under construction, changes in land use around the toll road have seen to be quite large at some points.

North and south of the Manado–Bitung Toll Road are directly facing the protected forest and water catchment areas, and are close to the active fault line. Therefore, the development of the surrounding area needs to be controlled to maintain the sustainability of the area from environmental problems and uncontrolled development.

To maintain the sustainability of the region, it is necessary to control the use of space. Ibrahim [7] argues that with spatial use control activities, identifications can be made at the same time to avoid the possibility of irregularities in spatial use. One of the efforts that can be made to control spatial use is by analyzing the carrying capacity and carrying capacity of the area so that the direction of development can be in accordance with the physical environment. Areas that have low-carrying capacity are directed as conservation areas. As for residential areas and other built-up lands, it is directed at areas with high land-carrying capacity.

2 Methods

The approach used in this research is a mixed-method, which is a research approach that involves two forms of research data, namely qualitative and quantitative [8]. This study uses a spatial analysis method to obtain maps of land capability units and land development capabilities, as well as the carrying capacity and regional carrying capacity by overlaying thematic maps according to the criteria that need to be met in each type of analysis. Descriptive and exploratory analysis was also carried out to determine and describe the direction of regional development in accordance with the potential problems of development around the Manado–Bitung Toll Road according to facts, analysis, literature review of previous research and information from informants.

3 Result and Discussions

3.1 Area Overview

The Manado–Bitung Toll Road stretches for approximately 39 km through the administrative areas of Manado City, North Minahasa Regency, and Bitung City. The Manado–Bitung Toll Road is planned to start operating at the end of 2019. This road is expected to support increased traffic on the Manado–Bitung route, supporting the tourism sector and economic growth in Manado, North Minahasa, and Bitung. This toll road will also be the main access road to the Bitung Special Economic Zone (KEK) and the Bitung International Hub Port to be built. Manado–Bitung Toll Road in the future can improve accessibility and connectivity, among them: (a) Facilitating traffic in developed areas; (b) Improving the distribution of goods and services to support economic growth; (c) Increasing the distribution of development and equity outcomes; (d) Relieving the burden on funds through the participation of road users.

In this study, the researcher limits the area of analysis in the research in the form of the lowest administrative boundary, namely the village boundary that is crossed by the Manado–Bitung Toll Road, and the village which is within a radius of about 1 km from the road body.

3.2 Land Capability Unit (LCU)

This analysis is carried out to obtain an overview of the level of land capability for regional development as a reference for direction of suitability and land use. This Land Capability Unit Analysis refers to the standards and criteria of the Minister of Public Works Regulation No. 20/2007 [9]. Based on the regulation, the determination of land capability is carried out by the superimpose method of various basic physical and environmental data including topographic and morphological data, soil type, rainfall, prone to natural disasters, land use, and others. The following are the weighting criteria for each type of LCU (Tables 1, 2, 3, 4, 5, 6, 7, 8 and 9; Fig. 1).

Table 1 Criteria and value of land capability unit (LCU) type: ease of work

Morphology	Slope (%)	Topography (masl)	Soil type	Landuse	Ease of work	Grade
Steep hills/mountains	>45	>3000	Regosol	Forest	Very low	1
Medium hills	25–45	2000–3000	Podsol, Andosol	Agriculture, plantation, dry land agriculture	Low	2
Low hills	15–25	1000–2000	Mediteran, Brown forest	Shrubs	Medium	3
Wavy	8–15	500–1000	Latosol	Dry land, heath	High	4
Plains	0–8	0–500	Alluvial	Settlement		5

Table 2 Criteria and value of land capability unit (LCU) type: foundation stability

Topography (masl)	Slope (%)	Morphology	Landslide vulnerability	Soil type	Foundation stability	Grade
>3000	>45	Steep hills/mountains	Zone 3 (high)	Regosol	Stability of the foundation is low	1
2000–3000	25–45	Medium hills	Zone 2 (medium)	Podsol, Andosol	Stability of the foundation is lacking	2
1000–2000	15–25	Low hills		Mediteran, Brown Forest		3
500–1000	8–15	Wavy	Zone 1 (low)	Latosol	Stability of the foundation is sufficient	4
<500	0–8	Plains		Alluvial		5

Table 3 Criteria and value of land capability unit (LCU) type: slope stability

Topography (masl)	Slope (%)	Morphology	Landslide vulnerability	Landuse	Slope stability	Grade
>3000	>45	Steep hills/mountains	Zone 3 (high)	Shrubs, fields, forests	Low	1
2000–3000	25–45	Medium hills	Zone 2 (medium)	Plantation, forest, grove	Less	2
1000–2000	15–25	Low hills		All	Medium	3
500–1000	8–15	Wavy	Zone 1 (low)	All	High	4
<500	0–8	Plains		All		5

Table 4 Criteria and value of land capability unit (LCU) type: water availability

Morphology	Slope (%)	Rainfall (mm/year)	Drought risk	Soil type	Landuse	Water availability	Grade
Steep hills/mountains	>45	1000–2000	High	Alluvial	Shrubs, fields, forests	Very low	1
Medium hills	25–45			Latosol	Gardens, forest, grove	Low	2
Low hills	15–25	2000–3000	Medium	Mediteran, Brown forest	All	Medium	3
Wavy	8–15			Podsol, Andosol	All	High	4
Plains	0–8	>3000	Low	Regosol	All		5

Table 5 Criteria and value of land capability unit (LCU) type: drainage capability

Topography (masl)	Morphology	Slope (%)	Rainfall (mm/year)	Soil type	Landuse	Drainage capability	Grade
<500	Steep hills/mountains	0–8	1000–2000	Latosol	Shrubs	Less	1
500–1000	Medium Hills	8–15			Forest		2
1000–2000	Low Hills	15–25	2000–3000	Mediteran, Brown forest	Agriculture, plantation	Enough	3
2000–3000	Wavy	25–45		Aluvial, Regosol	Dry land, heath	High	4
>3000	Plains	>45	>3000	Andosol	Settlement		5

Table 6 Criteria and value of land capability unit (LCU) type: erosion risk

Rainfall (mm/year)	Slope (%)	Morphology	Soil type	Landuse	Erosion risk	Grade
1000–2000	>45	Steep hills/mountains	Regosol	Grass, farm	High erosion	1
	25–45	Medium hills	Podsol Merah Kuning, Andosol	Plantation, bush	High enough erosion	2
2000–3000	15–25	Low hills	Meditera, Brown forest	All	Medium erosion	3
	8–15	Wavy	Latosol	All	Very low erosion	4
>3000	0–8	Plains	Alluvial	All	No erosion	5

Table 7 Criteria and value of land capability unit (LCU) type: morphological unit

Morphology	Slope (%)	LCU type: morphological unit	Grade
Steep hills/mountains	>45	Land capability of morphology is low	1
Medium hills	25–45	Land capability of morphology is lacking	2
Low hills	15–25	Land capability of morphology is moderate	3
Wavy	8–15	Land capability of morphology is enough	4
Plains	0–8	Land capability of morphology is high	5

3.3 *Land Development Capability (LDC)*

Development Capability Analysis is carried out to get an idea of the level of land capability that can be developed. The method used to create Land Development Capability (LDC) maps are overlaid on all LCU maps that have been made before. Then we need to multiply the value of each LCU maps with the weight value (Table 10) and sum up all of them, to get the LDC value.

$$\text{LDC Value} = \big[(\text{LCU } A \times \text{LCU Class Weight } A) + (\text{LCU } B \times \text{LCU Class Weight } B) \\ +(\text{LCU } C \times \text{LCU Class Weight } C) + (\text{LCU} \ldots \times \text{LCU Class Weight} \ldots) + \cdots\big] \tag{1}$$

The weight of each LCU shows the level of influence on the ability of land development. The following is the weighting for each LCU.

The classification of development capabilities can be seen in Table 11.

Based on the above analysis, they show that the research area is still dominated by a high level of development capability. This condition strongly supports the development of cultivation areas. More details about the distribution of land development capabilities in the Micro Area can be seen in Fig. 2.

3.4 *Carrying Capacity and Population Capacity*

Land-carrying capacity is an important thing that must be considered in regional spatial planning, in order to be able to support land use activities in a sustainable manner. Carrying capacity is the ability of the environment to support life while capacity is the ability of the environment to accommodate the population in it. The rapid development of the population around the Manado–Bitung Toll Road gives an indication of a serious urban problem. Therefore, we need an analysis of the carrying capacity and carrying capacity of land to determine the availability of effective land to support existing problematic land. Carrying capacity and population capacity are measured based on the calculation of the unit class of land capability that has been produced previously.

Table 8 Criteria and value of land capability unit (LCU) type: disaster risk

Slope (%)	Morphology	Topography (masl)	Landslide risk	Drought risk	Flood risk	Volcano risk	Earthquake risk	Landuse	Disaster risk	Grade
>45	Steep hills/mountains	>3000	High	High	High	High	High	Shrubs, field	High-risk disasters	1
25–45	Medium hills	2000–3000			Moderate	Moderate		Plantation, forest		2
15–25	Low hills	1000–2000	Moderate	Moderate			Moderate	All	Medium-risk disasters	3
8–15	Wavy	500–1000	Low	Low	Low	Low	Low	All	Low-risk disasters	4
0–8	Plains	<500			No risk	No risk		All		5

Table 9 Criteria and value of land capability unit (LCU) type: waste disposal capability

Slope (%)	Morphology	Topography (masl)	Landuse	Waste disposal capability	Grade
>45	Steep hills/mountains	>3000	Shrubs, fields	The ability of land for waste disposal is lacking	1
25–45	Medium hills	2000–3000	Plantation, thicket forest		2
15–25	Low hills	1000–2000	All	The ability of land for waste disposal is moderate	3
8–15	Wavy	500–1000	All	The ability of land for waste disposal is sufficient	4
0–8	Plains	<500	All		5

Areas with class E land capability are not able to be developed, so these areas are designated as protected areas with a 10% land cover ratio for activities that support protected areas. Class D has less capacity for development so this area is directed as a buffer zone with a 10% land cover ratio for activities that support buffer zones. Class C has moderate development capability, therefore, in this area development of cultivation areas can be carried out with limited developed land with a land cover ratio of 20%. Class B has sufficient development capability, in this area, the land cover ratio is directed at 30%. And class A has high development capability and in this area, it is very good to develop residential areas with a land cover ratio of 50% with an assumption of 30% in the form of green open space and 20% in the form of supporting facilities and utilities.

Land capacity is calculated with the assumption that one house consists of four family members which require a land area of 200 m^2. Thus, the population that can be accommodated in the area of land that can be developed divided by 200 m^2 divided by four people. The following are the results of the calculation of the carrying capacity and carrying capacity of the study area (Table 12).

3.5 *Regional Description and Development Recommendation*

The Manado–Bitung Toll Road is planned to have six interchanges as inlet/outlet, namely: (a) Manado Interchange Ring Road I which is at STA-0; (b) Sukur Interchange is at STA 7 + 200; (c) Air Madidi Interchange is at STA 12 + 600; (d) Cauditan Interchange is at STA 19 + 800; (e) Danowudu Interchange is at STA 28 + 900; and (f) Bitung Interchange is at STA 38 + 700. The six interchanges have different characteristics and development plans for the surrounding area. Figure 3 explains

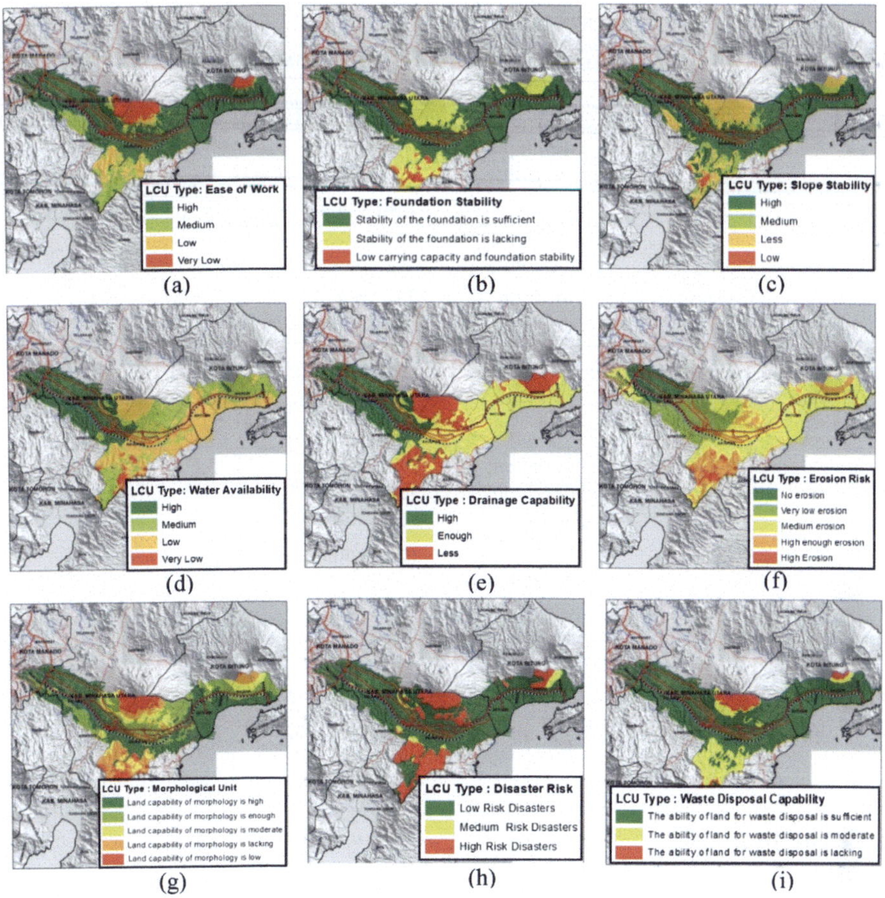

Fig. 1 **a** LCU ease of work; **b** LCU foundation stability; **c** LCU slope stability; **d** LCU of water availability; **e** LCU drainage capability; **f** LCU erosion risk; **g** LCU morphological unit; **h** LCU of disasters risk; **i** LCU of waste disposal capability

Table 10 LCU class weight value

Morphological	Weight = 5
Ease of work	Weight = 1
Slope stability	Weight = 5
Foundation stability	Weight = 3
Water availability	Weight = 5
Erosion risk	Weight = 3
Drainage capability	Weight = 5
Waste disposal capability	Weight = 1
Disasters risk	Weight = 5

Table 11 Land development capability

LDC value	Land capability	Development capability	Land use directive	Area (Ha)	Percent
32–58	E class	Low development capability	Conservation area	1.574	6
59–83	D class	Less development capability	Buffer zone	3.563	14
84–109	C class	Medium development capability	The area of cultivation, agriculture, plantations, with limited built-up land	3.690	14
110–134	B class	Sufficient development ability	Low-density settlement cultivation area	3.690	14
135–165	A class	High development capability	Residential area	13.025	51

the description of the regional development plan and the direction of development in accordance with the carrying capacity and carrying capacity of the region.

4 Conclusions

The construction of toll roads can trigger the growth of the region, especially around the area that is close to the inlet/outlet of the toll road, so that efforts are needed to control the use of space to match the carrying capacity and population capacity of the area. The development policy of an area needs to pay attention to the carrying capacity and population capacity of land; this has an effect on how much land can be developed. Development that is incompatible with its carrying capacity and population capacity can cause various environmental, social, and economic problems. Areas with low land-carrying capacity need special attention and policies need to be regulated so that this area is directed to be conserved, while the development of urban areas can be directed at areas that have high carrying capacity and population capacity value.

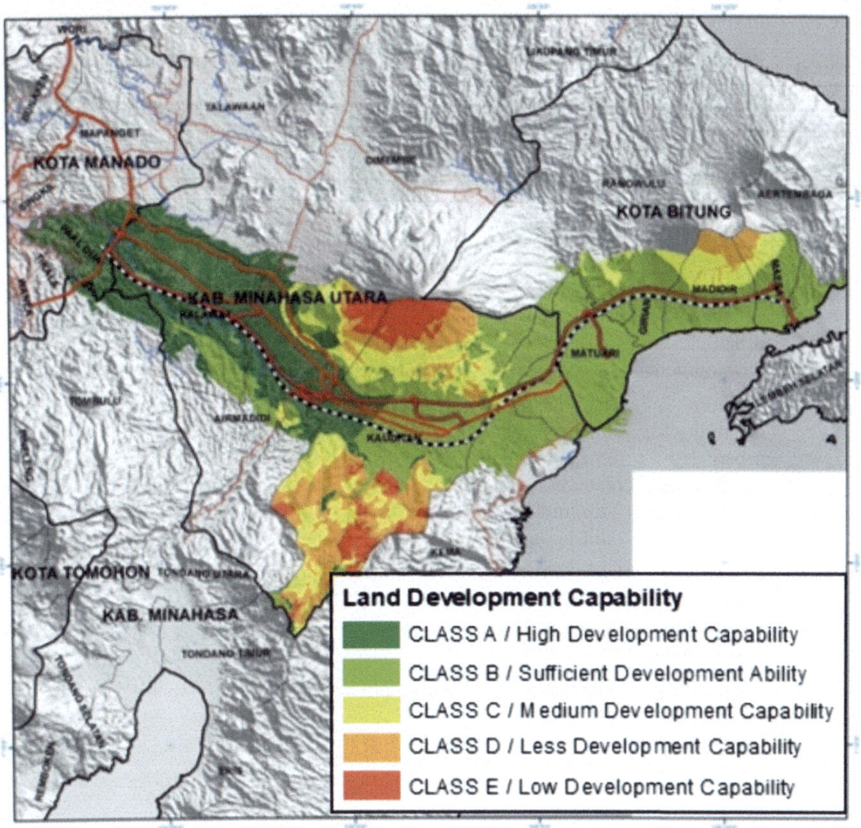

Fig. 2 Land development capability

Table 12 Carrying capacity and population capacity

Land capability	Development capability	Land-carrying capacity	Area (Ha)	Land cover ratio	Land that can be developed (Ha)	Population capacity (Person)
E class	Not able to be developed for urban areas	Protected area	1574	10% (Non-settlement)	–	–
D class	Less capability for urban development	Buffer zone	3563	10% (Non-settlement)	–	–
C class	Medium capability for urban development	The area of cultivation, agriculture, plantations, with limited built-up land	3690	20%	737,91	147,582
B class	Enough capabilities for urban development	Low-density settlement cultivation area, etc.	3690	30%	1,106,86	221,373
A class	High ability for urban development	Residential area	13,025	50%	651,240	1.302,480
Total					835,718	1,671,435

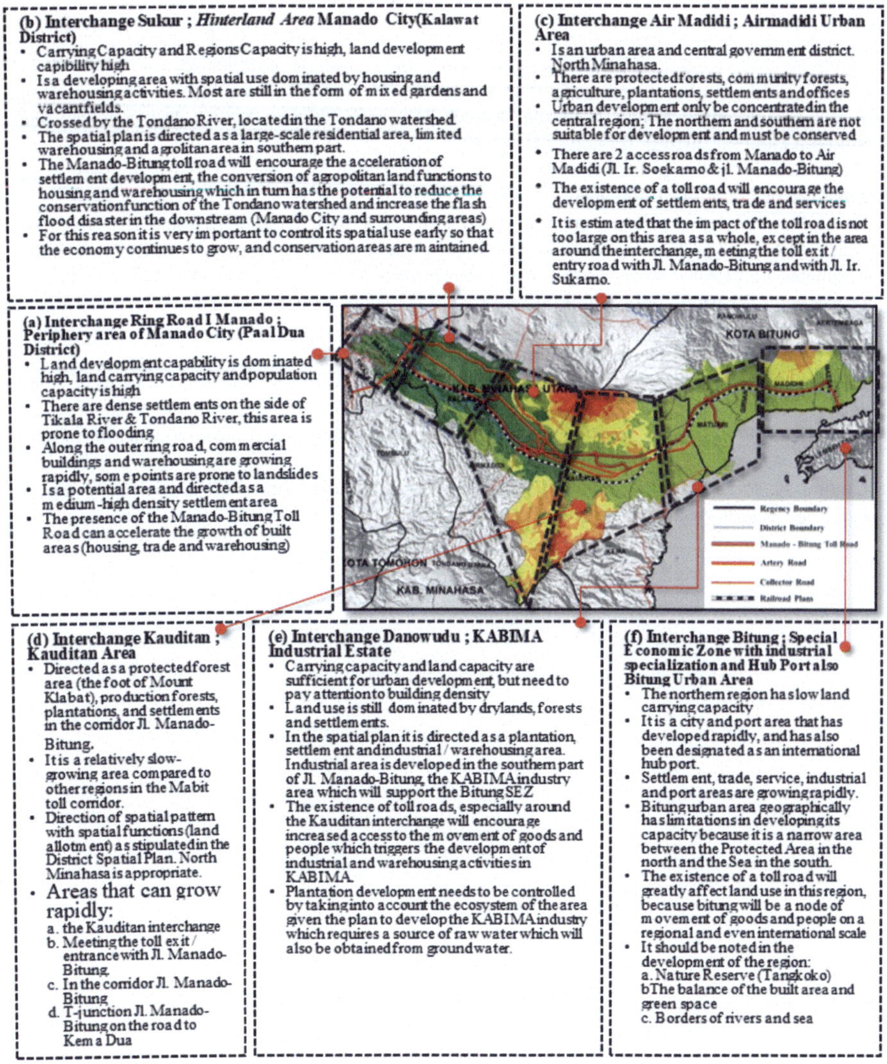

Fig. 3 Characteristics and direction of development of the area around the Manado–Bitung Toll Road

Acknowledgements This study is part of the Master's Indexed International Publication Study (PITMA B) 2019–2020, of the University of Indonesia contract number: NKB-1042/UN2.R3.1/HKP.05.00/2019, with the theme of "Practices of Planning and Conflict of Spatial Use in Urban Development".

References

1. Gie, K. K. (2002). Pembiayaan Pembangunan Infrastruktur dan Permukiman. Menteri Negara Perencanaan Pembangunan Nasional. Kepala Bappenas. Materi Kuliah Disampaikan Pada Studium General Institur Teknologi Bandung.
2. Haris, A. (2009). Pengaruh Penatagunaan Tanah Terhadap Keberhasilan Pembangunan Infrastruktur Dan Ekonomi. Bappenas.
3. Hurst, M. E. E. (1974). *Transportation geography*. New York: McGraw Hill.
4. Black, J. A. (1981). *Urban transport planning : Theory and practice*. London: Cromm Helm.
5. Edmonds, G. (1998). *Wasted time: The price of poor access*. Geneva: Development Policies Department, International Labour Office.
6. Dennis, R. (1998). *Rural transport and accessibility-a synthesis paper* (No. 1). Geneva: Development Policies Department, International Labour Office.
7. Ibrahim, S. (1998). Pengendalian Pemanfaatan Ruang Terpadu, Konsisten dan Berkualitas. *Jurnal Perencanaan Wilayah dan Kota, 9*(2).
8. Creswell, J. H. (2013). *Research design: Qualitative, quantitative, and mixed method*. United States of America: Sage Inc.
9. Umum, K. P. (2007). Peraturan Menteri Pekerjaan Umum Nomor 20 Tahun 2007 tentang Pedoman Teknik Analisis Aspek Fisik dan Lingkungan

Analysis of Gen Y Lifestyle Based on Life Cycle on Housing Preferences in Depok City

M. Akmal Farraz, Lita Sari Barus, Khoirunurrofik, and Joko Adianto

Abstract Gen Y is the age group with a range of 20–39 years and now is the largest generation in urban areas in Indonesia. As a result of their proportion to the urban population structure, the housing market has started to provide housing based on their preferences. This study aims to analyze the Gen Y lifestyle toward housing preferences with the scope of research in Depok City. Conducted with mixed methods, quantitative and qualitative research, the results of the study found that Gen Y who are married and are parents were the groups mostly affected by lifestyle when determining their housing preferences. The results of this study also showed that there is a lifestyle paradox between the two groups, 'Gen Y Junior' and 'Gen Y Senior,' which were affected by housing location and accessibility to educational facilities.

Keywords Gen Y lifestyle · Housing preferences · Urban area

1 Introduction

Demographic bonuses that are identical to the increasing age of the younger generation are happening in Indonesia. Based on data in 2017, the Gen Y population has reached 88 million or 33.75% of the total population. Also, the majority of this

M. Akmal Farraz · L. S. Barus (✉)
Urban Development Studies, School of Strategic and Global Studies, University of Indonesia, Depok, Indonesia
e-mail: litabarus@ui.ac.id

M. Akmal Farraz
e-mail: akmal.farraz@gmail.com

Khoirunurrofik
Department of Economics and Business, Faculty of Economics, University of Indonesia, Depok, Indonesia
e-mail: rofiq76@gmail.com

J. Adianto
Department of Architecture, Faculty of Engineering, University of Indonesia, Depok, Indonesia
e-mail: joko.adianto@gmail.com

S. S. Emamian et al. (eds.), *ICSDEMS 2019*,
https://doi.org/10.1007/978-981-15-3765-3_10

generation also lives in major cities in Indonesia such as Jakarta, Bandung, Medan, and Surabaya. As the dominant group, Gen Y who was born in the 1980s until the end of 2000 (aged 20–39 years) [1] who are entering a new phase of life such as fresh graduates, looking for work and young families will look for housing needs that significantly influence the housing market in the future.

The house as a means of life development will vary the desires of each individual or generation due to the influence of values, past experiences, and culture which is referred to as a lifestyle [2]. Lifestyle in housing studies is a popular concept that has been carried out by previous researchers such as Lee [3], Kwon et al. [4], and Rapoport [5]. Generally, these researchers use activities, interests, and opinions (AIO) in measuring lifestyle dimensions supported by social–demographic variables, namely gender, age, education, life cycle, transportation, etc., in order to obtain more accurate results Lee [3], Hoon Leh et al. [6], Nadiya [7], Berliner et al. [8].

Furthermore, in the housing study to determine individual preferences for housing, the researchers used housing attributes related to various aspects such as physical aspects, social elements, and the environment that associate housing attributes with location, type of house, accessibility, security, etc. [6, 9]. Important housing attributes are used to facilitate consumers in the decision-making process for housing choices [10] because of the consideration of various desires that are tailored to their daily activities. For example in the USA, Gen Y is more interested in crowded city life, residing in vertical housing, close to transportation routes and preferring to live in the city center because of a fast-paced lifestyle due to the influence of technology [11]. However, this is different in developing countries such as Malaysia and Indonesia that the younger generation still wants to live in the city center with large houses and close to work locations [6, 7].

This phenomenon illustrates that research on housing preferences cannot be generalized because it depends on the subject and location of the study. Moving on from these problems, this study focuses on Gen Y subjects who live in suburban areas, namely Depok City. The reason the researchers raised this theme is that the phenomenon of bonus demography that is identical to the younger generation also occurs in suburban areas such as Depok with 37.5 of its population being Gen Y which on the other hand the vision of Depok City is a residential city, especially for Jakarta residents [12]. In addition, the research subjects in Gen Y in the suburban areas were increasingly attractive and unique because of their lifestyle as commuter workers and the life cycle of unmarried people to have children who influenced their preferences toward home [10]. From these reviews, the purpose of this study was to analyze the characteristics and lifestyle of Gen Y in Depok City on housing preferences based on the life cycle.

2 Methods

This study uses two methods simultaneously, namely quantitative and qualitative methods (mixed methods). According to Creswell [13], mixed methods are two forms of mutually integrated methods in combining data, connecting data, and uniting data.

In this type of quantitative research, we used using a questionnaire with computer statistical tools such as Structural Equation Modeling (SEM), while the use of SEM is used to measure lifestyle factors using a 6-point scale, 1 (strongly disagree) to 6 (strongly agree).

To complete the results of research from quantitative methods, interviews with informants and several respondents were needed to collect complete information related to housing preferences. The results of the study are expected to be more comprehensive than the previous studies involving 400 respondents with accidental and stratified sampling techniques [14]. The reason the researchers chose this sampling technique was because they considered the ease of obtaining data while considering the representation of the population.

3 Result and Discussions

3.1 Profile of Sample

Based on field findings, the majority of the respondents were male sex totaling 63%. From the life cycle, respondents were unmarried by 42% and the smallest respondents were partners with 12% elementary school children. By age, based on stratified random sampling, respondents were represented in each age range with an average of 26%. In the socio-economic profile, the education level of the respondents is a bachelor (Strata 1) with a percentage of 47%. Of the types of work, the majority of respondents are employed as private employees with a percentage level of 37% followed by respondents who work as students with 19%, while the level of income of the majority of respondents have an average income of Rs. 5,520,000.

Overall, the respondents in this study were young urban middle class with unmarried status who began their careers as fresh graduates, working as private employees with an education level reaching Strata 1 and income levels above the UMR of Depok City of Rs. 5,520,000. Thus, the scheme of financing housing respondents is a group belonging to the middle class who is able to access financial assistance from banks.

3.2 Analysis of Lifestyle

In this study, the questionnaire was declared valid if the loading factor value was greater than 0.5 [15] and the hypothesis test is accepted if the *t*-scores were greater than 1.96 (>1.96) and the *p* value must have a 5% confidence level (<0.05).

In testing the validity of each question (Fig. 1), the activity variable ($X1$) of nine question items, the variable with the recreation indicator $X1.7$ with the question 'I enjoy free time by visiting places that are crowded by many people' is the best variable in shaping the activity variable with the highest variable loading factor value of 0.85. This means that the Gen Y lifestyle is more oriented to the activities of those who love to be outside home looking for entertainment.

Furthermore, the interest variable ($X2$) with holiday indicators ($X2.7$) with the question 'I like to follow the current trend' is the highest loading factor with a value of 0.81 which means that the Gen Y lifestyle is formed from a great interest in development current trends.

The best indicator in forming an opinion variable ($X3$) is itself in the variable ($X3.7$) with a loading factor of 0.87. With the question of the opinion of 'home showing self-image in the community,' it means that Gen Y in seeing home products does not only see it as a basic necessity, but a desired house can improve their self-image toward the community environment.

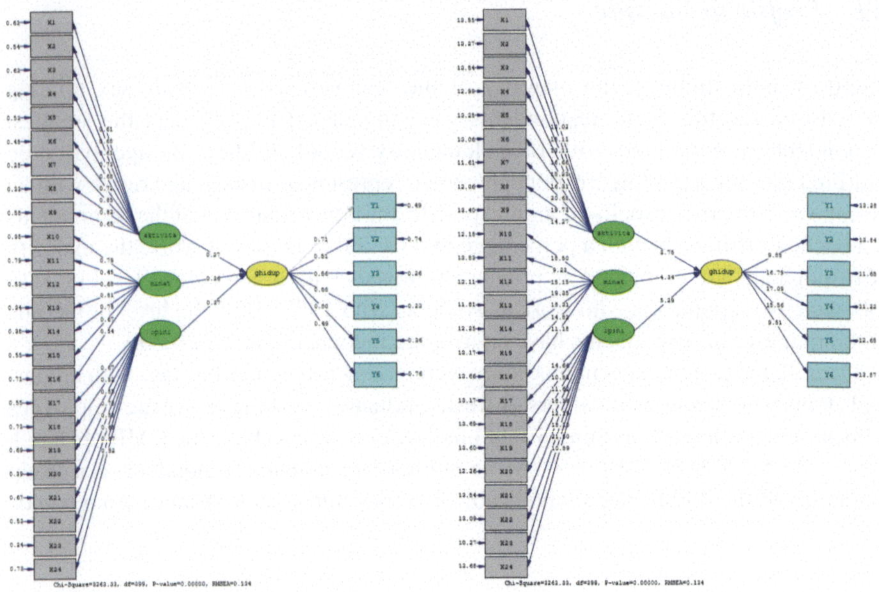

(a) (b)

Fig. 1 **a** Loading factor; **b** *T*-scores

The results of hypothesis testing show that the opinion variable has the highest *t*-score (5.29), which indicates that the higher the opinion that exists in Gen Y, the greater its influence in the formation of lifestyle in housing preferences.

After knowing the validity and reliability of the questionnaire, the next step is to classify lifestyle typologies. In this study, the researchers grouped six dimensions of lifestyle based on the question set in the questionnaire, namely community oriented (intimate relationship oriented to community or friends), traditional oriented (the desire for family as a priority and the tendency to become 'home people' which reduces their activities outside home), trendsetter oriented (interested in the latest products, activities outside the home, and the tendency to follow the latest trends), social prestige (consumption patterns aimed at showing self-image to the community), economic oriented (investment-oriented behavior, product quality, and hard work), and green living oriented (the desire for environmentally friendly living environment with good air circulation and perception of homes that have green open spaces).

Classification of lifestyle segments were based on the avarage calculation value [16]. Based on the calculation results show that Gen Y Junior has the characteristics of a green living oriented lifestyle (5.17) as a manifestation of the value of his life. While for Gen Y seniors, who generally have better careers and income, the orientation of their life values is social prestige (5.39). Furthermore, at the second highest average, value for Gen Y Junior is traditional oriented (4.69). Different conditions in Gen Y Senior are more oriented toward green living lifestyles (5.38).

On average sequential values in Gen Y Junior have a tendency toward economic-oriented (4.67), community-oriented (4.49), and trendsetter-oriented (4.36) lifestyles, and Gen Y Senior is more oriented toward the lifestyle of community oriented (5.08), traditional oriented (5.03), and trendsetter oriented (4.95) which is a reflection of lifestyle with the lowest average value.

3.3 *Housing Preferences Gen Y*

Based on literature review, this research uses housing attributes such as accessibility consisting of (proximity to work location, proximity to public transportation, proximity to religious facilities, proximity to school facilities, and proximity to near markets/malls), physical environment (free of disaster (e.g., flood) and atmosphere/calm environment, social environment (neighboring social interactions and environmental security), housing location (city center, suburbs, and villages), type of house (landed house and apartment), and ownership (own property, temporary renting, and renting for a long time).

From the results of research on housing preferences, between Gen Y Junior and Gen Y Senior, they have different desires (See Table 1). The results of this preference are explained by respondents (Gen Y Junior) that the desire to choose close to public transportation or work cannot be separated from their daily activities as commuters. In addition, the results of interviews with respondents who are generally under 30 years

Table 1 Comparison of housing preferences based on the life cycle

Respondent characteristics	Housing attributes					Total
(1)	(2)					(3)
Life cycle	Accessibility					
	Proximity to workplace	Proximity to public transportation	Proximity to religious facilities	Proximity to education facilities	Proximity to shopping facilities	
Gen Y junior (%)	36	37	8	11	8	100
Gen Y senior (%)	18	23	5	49	5	100
Physical environment						
	Free disaster		Calm environment			
Gen Y junior (%)	49		51			100
Gen Y senior (%)	49		51			100
Social environment						
	Environmental security		Neighborhood interaction			
Gen Y junior (%)	74		26			100
Gen Y senior (%)	66		34			100
Housing location						
	City center		Suburban		Rural	
Gen Y junior (%)	44		46		10	100
Gen Y senior (%)	55		23		22	100
Housing type						
	Landed housing			Apartment		
Gen Y junior (%)	93			7		100
Gen Y senior (%)	84			16		100
Mechanism of ownership						
	Own property	Temporary renting		Renting for a long time		
Gen Y junior (%)	98	2		0		100
Gen Y senior (%)	98	2		0		100

old show the desire to have a home in a suburban area because of a quieter and comfortable environment that is more suitable for a place to live than in a very crowded city center (Jakarta) and housing prices which are very expensive.

The desire for Gen Y 'Junior' homes tends to be free and not bound by external factors as experienced by Gen Y 'seniors' who generally have children. From the description of the interview, Gen Y 'Senior' toward their preferences at home tends to consider the interests of children today and their future such as the availability of educational facilities from the level of early childhood education to high school and a friendly environment for them to live.

The consideration of the distance between the house and the complete education facilities is an encouragement for young families in Depok City to determine the house with the type and location to be occupied by them in the future. Some of these considerations have significant implications for changes in their lifestyles such as household spending patterns, use of transportation (public or private) or type and location of work.

Specifically for Gen Y, 'Senior' chooses locations where they pay close attention to educational facilities for their children. In fact, the choice of home or city location (city center or suburban) is related to social policy support from the government as conducted by the Jakarta Government through the Jakarta Smart Card (KJP) which encourages many young families with low-income groups to choose to move from Depok goes to Jakarta as the center of the city even though it lives in a dense settlement and small houses.

3.4 Discussion

A person's lifestyle has is connected to their preferences for location and accessibility. From the results of studies that Gen Y 'Junior' with a green living orientation or traditional orientation have the desire to live in the suburbs is supported by public transportation. Gen Y Junior who is not married or has a family tends to have a traditional lifestyle that is prioritized by relationships and kinship with family, especially parents. This condition is due to career and income that is not yet stable for them own a house so that it requires the role of parents [11]. With this condition, Gen Y Junior tends to choose to live in suburban areas such as Depok rather than downtown (Jakarta) for a while so that economic and career conditions will be better and efforts to higher spending patterns if you have to live in Jakarta.

In the case of Gen Y Seniors, they prioritize the values of living with social prestige and green living and who are willing to live in the city center with more complete facilities specifically to meet the interests of children toward educational facilities supported by a better household economy and career.

In this case, the relationship between lifestyle and housing location can support what has been described by Rapoport [5] that those who are married will prioritize the interests of children in the choice of home location, but may differ from research by Pisman et al. [17] which indicates that although families will prefer to live in suburban areas because the environment is very comfortable.

Another important point is that Gen Y is eager to live in the city center because of more complete facilities and the availability of their jobs [18, 19] even at the risk of very expensive house prices and only getting small apartment rooms. But that does not mean that suburban areas are not as attractive as residential areas. There is also a part of this generation that has survived on the outskirts of the city for reasons of priority to the comfort and tranquility of life and the desire for a spacious home to accommodate nuclear family members or parents. But basically the preferences whether in the city center or in the suburbs would be very interesting if supported by integrated public transportation and educational facilities that meet the needs of children in the future. For this reason, it is very important to know housing preferences in this generation to improve their quality of life and the sustainability of a city, especially in the provision of decent housing.

4 Conclusions

The results of this study involving 400 respondents illustrate that the characteristics of Gen Y in Depok City are urban middle class who are generally unmarried with income levels above the Depok City Minimum Wage. The majority of respondents are men because of housing needs in patrilineal cultural characteristics such as Indonesia housing is a responsibility that is generally charged to male gender groups.

From the dimensions of lifestyle analysis, Gen Y in Depok City is a group with a life orientation toward family and environment. They are a group that has an orientation to gather and interact with family. The interesting thing about the comparison of life cycles based on the life cycle is the Gen Y Junior lifestyle phenomenon (not yet having children) with an orientation to the traditional lifestyle of prioritizing family. Against housing preferences, Gen Y 'Junior' prefers to be close to public transportation and prefers to live in the suburbs. This condition is inversely proportional to Gen Y Senior (already having children) who tends to be socially oriented prestige with social status as a life value. To meet their needs, Gen Y Senior prefers to live in urban areas with complete facilities, especially those close to educational facilities to fulfill children's interests as the main thing.

Acknowledgements This article is part of Grant of International Publications for Students Final Project of University of Indonesia.

References

1. Badan Pusat Statistik Indonesia. (2018). Profile of the millennial Generation of Indonesia.

2. Jansen, S. J. T., et al. (2011). *The measurement and analysis of housing preference and choice.* London, New York: Springer, Dordrecht Heidelberg.
3. Lee, H. J. (2005). *Influence of lifestyle on housing preferences of multifamily housing residents* (Dissertation). Virginia Polytechnic Institute and State University, Virginia.
4. Kwon, H. J., et.al. (2016). U.S. Boomers' lifestyle and residential preferences for later life. *Journal of Asian Architecture and Building Engineering, 15*(2), 255–262.
5. Rapoport, A. (1980). *Human aspects of urban form (towards a man-environment approach to urban form and design).* Oxford: Pergamon Press.
6. Hoon Leh, O. L., et al. (2016). The housing preference of young people in Malaysian urban areas: A case study Subang Jaya, Selangor. *Malaysian Journal of Society and Space, 12*(7), 60–74.
7. Nadiya, E. (2017). Preference study of Y generation in choosing residential in West Jakarta. *Jurnal Muara Ilmu Ekonomi dan Bisnis, 1*(1), 145–152.
8. Berliner, et.al. (2015). *Generation Y: Lifestyles and mobility choices of millennials in California, and the motivations behind them.* Presentation Series Institute of Transportation Studies. University of California, Davis.
9. Wu, F. (2010). Housing environment preference of young consumers in Guangzhou, China. *Journal of Property Management, 28*(3), 174–192.
10. Yin, Y. (2015). *Lifecycle housing preferences of middle class residents in Guangzhou: An analytic hierarchy process (AHP) approach* (Thesis). University of Hong Kong, Hong Kong.
11. Lachman, M. L., & Deborah, L. B. (2015). *Generation Y: America's new housing wave.* Urban Land Institute: Washington, D.C.
12. Badan Pusat Statistik Kota Depok. (2018). Depok in 2017 Numbers.
13. Creswell, J. W. (2014). *Research design: Qualitative, quantitative, and mixed methods approach* (4th ed.). London: Sage Publications.
14. Sugiyono. (2018). *Quantitative, qualitative, and R&D research methods.* Bandung: Alfabeta.
15. Ghozali, I. (2017). *Structural equation model: Concepts and applications with Amos 24 Bayesian update SEM* (7th ed.). Semarang: Badan Penerbit Universitas Diponegoro.
16. Sudjana. (2001). *Metode & Teknik Pembelajaran Partisipatif*. Bandung: Falah Production.
17. Pisman, A., et al. (2011). Urban and suburban lifestyle and residential preferences in a highly urbanized society. Case study in Ghent (Flanders, Belgium). *Belgeo, 1–2,* 89–104.
18. Farraz, M. A., & Barus, L. S. (2019). *Housing preferences and choice young families commuters in Depok City* (Vol. 264). Indonesia: Earth and Environmental Science (EES).
19. Axford, N., et.al. (2016). Millenials myths and realities. Global Overview, CBRE Research.

Architect's Knowledge and Perception to Apply Green Building Aspect in Design

Togu Riotama, Andrean Eka Lucianto, and Herdis Herdiansyah

Abstract Buildings contribute to environmental degradation due to large energy usage and carbon emission production. Awareness to protect the environment, and reduce the impact of development, make the concept of sustainable development is applied through good planning by implementing sustainable design. As a planner of a building or even a region, architect needs to implement a sustainable design to reduce the damage caused by development. This study aims to analyze the awareness of the architect in applying sustainable design as an effort to protect the environment. The study was conducted using quantitative methods, by collecting data through questionnaires. The sampling technique is carried out by purposive sampling method. Purposive sampling is a technique for sampling data sources with certain considerations. The results have shown that there are influence between environmental awareness and green building knowledge, toward architect perception on green building implementation.

Keywords Green building · Architect · Knowledge · Perception

1 Introduction

The increasing of global temperature or commonly called global warming is caused by emissions from industries, transportation, and buildings [1]. However, building and construction industries produced about 39% of CO_2 emissions globally. This is the largest portion compared to transportation, 28%, and other industries 31% [2].

Building is needed by humans to contain their activities such as living, working, or recreating. However, building also contributes to environmental damage, not only when the building is under process, but also when it operates. Based on USGBC in 2008, buildings consume a lot of materials, including woods, and use resources such as water and energy. Buildings also produces about 40% of US non-industrial waste

T. Riotama · A. E. Lucianto · H. Herdiansyah (✉)
School of Environmental Science, Universitas Indonesia, Jl. Salemba Raya no. 4 Central, Jakarta 10430, Indonesia
e-mail: herdis@ui.ac.id

S. S. Emamian et al. (eds.), *ICSDEMS 2019*,
https://doi.org/10.1007/978-981-15-3765-3_11

and contribute to global warming as they create 36% of carbon dioxide emissions [3].

With the awareness to protect the environment and reduce the impact of development, make the concept of sustainable development is applied more, including in the construction industry sector. Green building is one of the sustainable development movements in construction industry.

Green building concept was developed in 1990 in order to implement energy-efficient building design guidelines and eco-friendly renewable materials so environmental damage could be reduced [4]. Green building assessment tool (GBAT) was created to see if a building meets the requirement to be a green building. This tool also became a guideline for architects to design sustainable buildings.

The GBAT was developed by green building council, an independent organization that is concerned in green building practice. Green building council exists in some countries, including Indonesia. Green Building Council Indonesia (GBCI) was formed by experts in design and construction industry in 2009. This council focuses in socialization and transformation of sustainable green principles, especially in building construction industry in Indonesia [5]. GBCI developed Greenship, a rating tool to assess green building according to condition and situation in Indonesia.

To create a sustainable building, the green building criteria need to be considered from the early stages of the design process [6]. Therefore, experts in building industry have to consider green building strategies and sustainability principles in building development [1].

Architects need to respond to the environmental issues and integrate design process to create a sustainable building as the market demands nowadays [7–9]. Therefore, it is important for architects to have environmental awareness and green building knowledge, so they could provide sustainable design to reduce environmental damage that happens in building process and when the building operates. This study aims to analyze environmental awareness and green building knowledge related to architect perception in applying green building aspect in their design.

This present study is unique compared to existing literature because of several reasons. Firstly, it discusses green building issue from architect's perception and knowledge, while the previous studies criticize green building from users or public perspectives [10–12]. This study helps to understand architect's perception of designing green buildings.

Secondly, this research discusses architect's knowledge about green building as a whole. Existing literature commonly focus on partial aspects of green building, such as building envelope [13] and wood materials [3, 14]. Therefore, this research helps to give a comprehensive architect's perspective about green building.

2 Methods

2.1 Green Building

The research of United States Green Building Council Research Committee in 2008 revealed that buildings have big contribution to environmental damage. A big amount of energy is used for lighting and mostly for air conditioning when the building operates. Green building refers to buildings that avoid or reduce negative impacts to the environment [15].

Green building has become popular nowadays, since many green building councils have been established in many countries. There are around 70 councils registered as members of World Green Building Council (WGBC). These councils focus on green building issues including building certification. A building is recognized as a green building by assessment process using green building rating tools (GBRT). There are some GBRT that have been developed in the world, for example LEED, BREEAM, CASBEE, and Green-star [16]. These GBRT have different criteria and credits for assessment. The GBRT used in one country might be different from another country, and it depends on the environmental, social, and economic conditions. Therefore, green building council might develop its own tools to adjust with the condition in its country.

In 2009, some professionals in design and construction industry who are concerned about green building practices formed Green Building Council Indonesia (GBCI). GBCI focuses in green building implementation and has developed a rating tool for assessing green building in Indonesia, named greenship. This rating tool consists of six assessing categories: Appropriate Site Development (ASD); Energy Efficiency and Conservation (EEC); Water Conservation (WAC); Materials & Resources Cycle (MRC); Air Quality and Leisure Air (Water Indoor Health and Comfort/IHC); and Building and Environment Management (BEM) [4].

2.2 Data Collection

This research conducted a survey questionnaire to analyze the relation between environmental awareness and green building knowledge toward perception of green building concept. The questionnaire was designed using Google Forms and distributed to architects from different firms and ages in Jakarta. The questions used short answer questions and Likert Scale-type questions with five alternative answers: Strongly Agree, Agree, Neutral, Strongly Disagree, and Disagree. This questionnaire aims to understand the architect's environmental awareness and green building aspects knowledge.

The questionnaire consisted of four sections. Section 1 contains questions about respondent's general data, such as name, age, and working experience. Section 2 dealt with respondent's awareness about some environmental issues. Section 3 was about respondent's basic knowledge of green building. Section 4 was about respondent's perception of green building.

Data processing was conducted to see the relation between variable environmental awareness and green building knowledge toward the green building perception. The data was processed using SPSS software with multiple linear regression method. Multiple linear regression is a regression that has one dependent variable and more than one independent variables [17]. The multiple linear regression equation models are as follows:

$$Y = a + b_1 X_1 + b_2 X_2 + \cdots +$$

Y is the dependent variable
X_1 is the first independent variable
X_2 is the second independent variable
a, b_1, b_2, e are the constants.

3 Results and Discussion

There is fifty-six (56) data from questionnaires that were returned out of the eighty (80) questionnaires administered. The data was presented using descriptive statistical tools. The respondents were mostly architects under 40 years (73%).

The working experience showed that most respondents (41%) have 0–5 years of working experience, while there are 21% respondents with 6–10 years of working experience, 11% respondents with 11–15 years of working experience, and 27% respondents with 16–20 years of working experience.

3.1 Environmental Awareness

Table 1 shows the questionnaire results on environmental awareness section. Most respondents, 48%, strongly agree that human activities contribute to environmental damage, while the others 40% agreed, and the others neutral. The respondents that strongly agreed to sort out the type of garbage before disposing it is 54 and 28% disagreed, while the other 18% respondents were neutral. The percentage of respondents that strongly agreed to avoid using plastic were 30%, while 23% agreed, 34% neutral, 11% disagreed, and 2% strongly disagreed. The amount of respondents that strongly agreed to use recycled paper at work were 62%, while 20% agreed, 11% neutral, 5% disagreed, and the other 2% strongly disagreed.

Table 1 Environmental awareness result

	Human activities contribute to environmental damage (%)	I sort out the type of garbage that I disposed of (%)	I avoid using plastic by carrying shopping bag (%)	I use recycle paper at work (%)
SA	48	54	30	62
A	41	28	23	20
N	11	18	34	11
D	0	0	11	5
SD	0	0	2	2
Total	100	100	100	100

SA Strongly agree; *A* Agree; *N* Neutral; *D* Disagree; *SD* Strongly disagree

3.2 Green Building Knowledge

Green building knowledge section questionnaire consists of four general questions about green buildings. The first question was about the main aspect of green building according to them. The results were grouped into some categories based on the main idea of respondent answer. Table 2 shows that most of the respondents (42%) consider minimum energy usage as the main aspect of a green building.

In order to know the architect's basic knowledge about green building in Indonesia, there are some questions with multiple choices. Table 3 shows that all respondents knew about the green building institution in Indonesia. Most of respondents (89%) knew about the green building rating tools in Indonesia, and 52% of respondents knew about one category in that GBRT.

Table 2 Green building aspects

Aspect of green building	Results (%)
Sustainable	35
Minimum energy usage	42
Low carbon footprint	8
Eco-friendly material	15
Total	100

Table 3 Green building knowledge

	Name of institution that focus on green building in Indonesia (%)	Name of green building rating tools in Indonesia (%)	Name of one category in rating tools in Indonesia (%)
True	100	89	52
False	0	11	48
Total	100	100	100

Table 4 Green building perception result

	GB aspect are easily applied in the design (%)	The application of GB aspect tends to be cheap (%)	GB aspect are important in the design (%)	I applied GB aspect to the design (%)
SA	20	23	71	23
A	41	36	18	56
N	29	12	0	16
D	5	18	11	5
SD	5	11	0	0
Total	100	100	100	100

SA Strongly agree; *A* Agree; *N* Neutral; *D* Disagree; *SD* Strongly disagree

3.3 Green Building Perception

Table 4 shows the questionnaire results on green building perception section. The results show that 41% of the respondents agreed that green building aspect is easily applied in the design. The other 20% were strongly agree, while 29% were neutral, 5% were disagree, and the other 5% were strongly disagree. The respondent that strongly agree that the application of green building aspect tends to be cheap were 23%, and 36% were agree, while the other 12% respondents were neutral, 18% were disagree, and 11% were strongly disagree. The percentage of respondents that strongly agreed that green building aspects are important were 71%, while 18% were agree, and the other 11% were disagree. Respondents that strongly agreed to apply green building aspect to design were 23%, while 56% agreed. The other 16% were neutral, and 5% were disagree.

3.4 Data Analysis

To find out the relationship between variables, data processing was performed using SPSS software with multiple linear regression method. The results can be seen in Table 5.

From the results of the analysis, the value of sig <0.05, then there is an influence between the independent variables on the dependent variable. Thus, an increase in environmental awareness and green building knowledge can enhance the perception to apply green building in design.

The main barrier in green building application is lack of information, knowledge, and awareness of the stakeholders, including architects [18]. Therefore, it is important for government and institutions that focus on green building, for example GBCI, to promote green building application. According to Bhumkar [8] architectural education needs to include subject about environmental issues to increase student awareness in implementing sustainable architecture design. Similarly, Wyckmans

Table 5 Multiple linear regression analysis result

Coefficients[a]

Model		Unstandardized coefficients		Standardized coefficients	*t*	Sig.
		B	Std. error	Beta		
1	(Constant)	7.209	2.579		2.796	0.007
	Environmental awareness	0.398	0.150	0.340	2.645	0.011
	Green building knowledge	0.282	0.124	0.324	2.518	0.021

[a]Dependent variable: green building perception

[19] stated that universities and architectural schools must play an active role in educating about environmental issues.

The growth of green buildings in Indonesia needs to be increased to reduce the negative impact to environment. Architects as building planners that are involved in development process need to have environmental awareness and knowledge in applying green building aspects in the design. Through environmental education for architects and architecture students, it is expected that will enhance their perception in implementing sustainable design through green building aspects.

Environmental education could be included in architect training or expertise upgrading program in order to get their license. This is expected to increase environmental insight that could encourage architects to implement sustainable designs.

4 Conclusions

The results showed that environmental awareness and knowledge about green building had an influence on the perception of architects to apply the green building aspects of design. However, lack of awareness, knowledge, and information are the main obstacles in green building aspects implementation. Therefore, increasing environmental awareness and knowledge of green building need to be improved through formal education in universities or architecture schools and training institutions, for example GBCI. Thus, architects are expected to be more active in applying the green building aspects in design.

Acknowledgements This research is funded by Directorate General Higher Education (DIKTI), Ministry of Research, Technology and Higher Education of the Republic of Indonesia. (contract number NKB-1903/UN2.R3.1/HKP.05.00/2019).

References

1. Isopescu, D. N. (2018). *IOP Conference Series: Materials Science Engineering* (Vol. 399) 012026
2. UN Environment and International Energy Agency. (2017). *Towards a zero-emission, efficient, and resilient buildings and construction sector* (Global Status Report 2017).
3. Hoffman, A. J., & Henn, R. (2008). Overcoming the social and psychological barriers to green building. *Organization and Environment, 21*(4), 390–419.
4. Bowers, T., Ganguly, I., & Eastin, I. (2014). Eco-labeled wood products in the U.S. residential construction industry: Architects' awareness and usage of certified wood and green building programs. *The Forestry Chronicle, 90*(05), 605–613. https://doi.org/10.5558/tfc2014-124
5. Green Building Council Indonesia. (2019). *About GBC Indonesia.* Retrieved from http://gbcindonesia.org/.
6. Farid, V. L., & Wonorahardjo, S. (2018). *IOP Conference Series: Earth Environment Science* (Vol. 152) 012012.
7. Altomonte, S. (2009). Environmental education for sustainable architecture. *Review of European Studies, 1*(2). https://doi.org/10.5539/res.v1n2p12.
8. Bhumkar, M., & Dhavale, S. A. (2017). *Environmental Awareness and Architectural Pedagogy, 10*(1), 201–205.
9. Bonenberg, W., & Kapliński, O. (2018). The architect and the paradigms of sustainable development: A review of dilemmas. *Sustainability, 10*(2), 100. https://doi.org/10.3390/su10010100.
10. Sabar, R., Anuar, N. K., Mutazam, M., Nifa, F. A. A., & Lin, C. K. (2018). Awareness of green building design in Malaysia. In *AIP Conference Proceedings*, 2016. https://doi.org/10.1063/1.5055528.
11. Aminrad, Z., Sayed Zakariya, S. Z. B., Samad Hadi, A., & Sakari, M. (2013). Relationship between awareness, knowledge and attitudes towards environmental education among secondary school students in Malaysia. *World Applied Sciences Journal, 22*(9), 1326–1333. https://doi.org/10.5829/idosi.wasj.2013.22.09.275.
12. Yusoff, M. N., Nawi, M. N., & Ibrahim, S. (2015). The study of green building application awareness. *Jurnal Teknologi, 75*(9). https://doi.org/10.11113/jt.v75.5228.
13. Akinola, A. O., Adeboye, A. B., Oluwatayo, A., Alagbe, O., Babalola, O., & Afolabi, A. O. (2018). Survey dataset on architect's awareness and adoption of building envelope technologies for energy efficient housing in Lagos State. *Data in Brief, 19,* 1894–1901. https://doi.org/10.1016/j.dib.2018.06.093.
14. Markström, E., Kuzman, M. K., Bystedt, A., Sandberg, D., & Fredriksson, M. (2018). Swedish architects view of engineered wood products in buildings. *Journal of Cleaner Production, 181,* 33–41. https://doi.org/10.1016/j.jclepro.2018.01.216.
15. World Green Building Council. (2019). *About green building.* Retrieved from https://www.worldgbc.org/what-green-building.
16. Olakitan, J., & Ayşe, A. (2018). Social criteria of sustainable development in relation to green building assessment tools. *Environment, Development and Sustainability.* https://doi.org/10.1007/s10668-018-0184-1.
17. Sujarweni, V. W. (2014). *SPSS for research* (available in Bahasa Indonesia only). Yogyakarta: Pustaka Baru Press.
18. Darko, A., & Chan, A. P. C. (2017). Review of Barriers to Green Building Adoption. *Sustainable Development, 25*(3), 167–179. https://doi.org/10.1002/sd.1651.
19. Wyckmans, A. (2008). Environmental learning in architecture from individual choice, pp. 73–87

Analysis of the Impact of Urban Development on River Water Quality Case Study of the Pesanggrahan River

Muslim Aminuddin and Chotib

Abstract The Pesanggrahan River from its width characteristics is an intermediate river. The chemical and biological contents of Pesanggrahan River water show that the Pesanggrahan River has been polluted. Water pollution in the Pesanggrahan River is greater in the downstream area, which is due to the accumulation of chemical compounds from industrial and domestic waste. Most of the Pesanggrahan Watershed are residential areas. City development in the Pesanggrahan Watershed has a major influence on the decline in the water quality of the Pesanggrahan River. The biggest development occurred in the period of 2004–2010. Then, in the 2010–2013 period, the development was more on the changes in the structure of the Pesanggrahan River flow, namely on river widening and straightening. The area in the Pesanggrahan Watershed is widely used as an illegal industrial area, thus, violating existing spatial provisions. The role of the Jakarta City Government in maintaining river water quality is in the function of building and supervising buildings that violate the rules. This refers to the granting of permits and finally to prosecution of parties who violate and play a role in decreasing the quality of the Pesanggrahan River water.

Keywords Pesanggrahan River · Pesanggrahan watershed · Water quality

1 Introduction

The problems caused by pollution that occur in the Pesanggrahan River cannot be separated from the role of buildings built in the area. Therefore, it is necessary to identify water quality which will later be linked to the pattern of urban development in the Pesanggrahan River. The process of identifying water quality is done by using the Water Quality Index (WQI) calculation found in the Pesanggrahan River. The indicators observed in the Water Quality Index are [4]: water temperature, pH, dissolved oxygen (DO), total dissolved solids (TDS), total suspended solids (TSS), water hardness, nitrate (NO_3), nitrite (NO_2^-), biochemical oxygen demand (BOD),

M. Aminuddin (✉) · Chotib
Urban Development Study Program, University of Indonesia, Kampus UI Salemba, Jakarta, Indonesia
e-mail: muslim.aminuddin1993@gmail.com

S. S. Emamian et al. (eds.), *ICSDEMS 2019*,
https://doi.org/10.1007/978-981-15-3765-3_12

chemical oxygen demand as O_2 (COD), Fecal Coliform Bacteria, and Total Coliform Bacteria.

Water Quality Index is used by many developing countries because it has analytical costs involved, which can be a limiting factor for water quality assessment with a small budget [6]. Water Quality Index in this study is used as a reference to determine the effect of development around the Watershed Pesanggrahan on environmental degradation in water. So that in the future the DKI Jakarta Government can take steps in regulating urban development and returning the function of the Pesanggrahan River to its original function by considering the quality of water available at the Watershed, especially the Pesanggrahan River.

2 Materials and Methods

This research was conducted using qualitative and quantitative approaches using field survey methods. The survey method is an investigation conducted to obtain facts from the symptoms that exist and look for facts. The survey method dissects and skins and recognizes problems as well as obtaining justification for the conditions and practices that are taking place. The survey method also evaluates and compares things that people have done in dealing with similar situations or problems, and the results can be used in making plans and making decisions in the future [8].

Sampling was carried out along the Pesanggrahan River. The distance between one sampling point and another point is about 500–700 m. The distance was chosen so that data on the chemical and biological compounds that are present in the Pesanggrahan River water can be clearly known. Therefore, the total samples obtained in this river study are 37 observation points knowing that the length of the Pesanggrahan River in the urban area of DKI Jakarta is ±20 km.

Pesanggrahan River water sampling is carried out in the Pesanggrahan River Watershed (DAS) area which covers four districts; Pesanggrahan District, Kebayoran Lama District, Kebon Jeruk District, and Kembangan District. Actually, the urban headwaters of the Pesanggrahan River are in the Cilandak area, but Watershed, which is located in that area, is part of the entry area in the Tangerang Regency, Banten Province.

The location of this study was chosen because the Watershed Pesanggrahan has urban characteristics and the Pesanggrahan River flows through dense residential areas.

3 Result and Discussion

3.1 *Physical Analysis*

Pesanggrahan River has a complex system in the urban water system in DKI Jakarta. Judging from the classification described by Kern (1994), Pesanggrahan River is part of the medium river classification. In the process of field observations, the width of the Pesanggrahan River ranges from 10 to 20 m in almost all parts of the Pesanggrahan River, except at points 24 and 37, which are tributary branches which have a width of 20–40 m.

A river that is in a branching with a tributary usually has a river width that is relatively larger compared to other watersheds. This is caused by several factors such as erosion caused by the flow of the river that erodes the river border or river banks.

The regularity of the Pesanggrahan River flow, according to the classification described by Rosgen (1996) in Maryono [6], is a straight-type river; this can be seen with the Pesanggrahan River flow pattern that is described and viewed using the help of the Google Earth application and also a map of the Regional Spatial Plan DKI Jakarta which describes the Pesanggrahan River is in type G with a slope of 0–2%. The slope is usually located in the Watershed area which is upstream of the river. So it has other characteristics, namely the river flow velocity which tends to be slower compared to the downstream areas of the Pesanggrahan River in the Bogor and Depok regions.

3.2 *Water Quality Index of the Pesanggrahan River*

3.2.1 Temperature

The highest value of the observation of the water temperature of the Pesanggrahan River is at observation point 27, with a temperature value of 31 °C. The meaning is not clear consider revising, and the lowest value is at observation point 1 or DKI Jakarta's upstream urban Pesanggrahan River with a temperature of 28 °C and the low value can be caused by weather factors where at 1–8 observations are observed after rain. The average temperature in the study of the Pesanggrahan River is 29.45 °C.

Calculation of the temperature value is very difficult to identify the cause, because many factors affect the temperature value of water [3]. Besides the influence of nature factor such as rain and heat weather can also affect water temperature. However, if the weather conditions are relatively stable, the existing conditions of the Watershed Pesanggrahan can be identified, whether the Watershed Pesanggrahan is an area that has an area with good ecosystem conditions or not. The influence of ecosystems such as trees and plants can cause water shad to be a low water surface temperature value. Conversely, if the condition of the Watershed Pesanggrahan does not exist or a bad ecosystem is found/there are no trees, then the surface temperature value of the Pesanggrahan River water tends to follow the existing air temperature. The influence of basic chemical compounds (usually in the form of detergents) can theoretically cause the condition of the relative temperature of the water to be higher than the average value of the existing air temperature.

3.2.2 pH

The highest value of the observation results of the water pH of the Pesanggrahan River is at observation point 37 which is the observation point area downstream of the Pesanggrahan River with a pH value of 8.70 or having alkaline properties, and the lowest pH value of the water of the Pesanggrahan River is at observation point 1 or urban DKI Jakarta Sungai Pesanggrahan with a pH of 7.2 having normal alkaline properties. The average pH in the study of the Pesanggrahan River is pH 7.84. The increase in acidity (pH), which has a tendency to increase, can be ascertained due to the increasing number of chemical compounds with alkaline content that is discharged into the Pesanggrahan River at each point downstream of the Pesanggrahan River. These basic chemical compounds can all be found in detergents or cleaning agents, both from domestic and industrial waste products. pH is a chemical property that is influenced by the content/mixing of a substance's properties in proportion to volume.

3.2.3 Dissolved Oxygen

Increased diffusion of oxygen or dissolved oxygen from the air into the waters is usually caused by wind, falling rain to the surface, temperature, pressure, and the content of various dissolved ions that can affect the level of dissolved oxygen. The following are the results of the measurement of dissolved oxygen in the water of the Pesanggrahan River.

The highest value from the observation of dissolved oxygen (DO) in Pesanggrahan River water is at observation points 1, 2, 3, and 5, namely in the upstream urban observation point areas of DKI Jakarta Pesanggrahan River with a DO value of 3.50 mg/L. This point can be said to be the cleanest point and has a Watershed (DAS) tendency with a fairly good ecosystem, and the lowest DO value of Pesanggrahan River water is at observation points 36 and 37, namely the downstream observation points of the Pesanggrahan River with DO value of 2.00 mg/L. Watershed at this point tends not to have a good ecosystem with the existing condition of the Pesanggrahan River floodgate, where the garbage is carried by the Pesanggrahan River, or transported or accommodated (the existing condition at point 36 is a flood gate and a "taman kota" landfill, kembangan). The average value of Dissolve Oxygen in the Pesanggrahan River water study is 2.75 mg/L.

The low value of dissolved oxygen in river water is caused by accumulated garbage, a reduction in plant vegetation that is always decreasing in the Watershed

Pesanggrahan and waste factor that is discharged directly in the Pesanggrahan River. In addition, the low value of dissolved oxygen (DO) proves that the Pesanggrahan River has been polluted and the aquatic organisms that live in it are disturbed or in other words aquatic organisms (for example: fish and bacteria) in the Pesanggrahan River flow decline, which is directly proportional to the declined oxygen content [9].

3.2.4 Biochemical Oxygen Demand (BOD) and Chemical Oxygen Demand (COD)

The highest value of the observation of biochemical oxygen demand (BOD) of Pesanggrahan River water is at observation point 26, which is the observation point of the Permata Srengseng complex with the biochemical oxygen demand value of 5.50 mg/L and the lowest value of the biochemical oxygen demand of Pesanggrahan River water is at the observation point 1, namely the DKI Jakarta upstream urban observation point area Pesanggrahan River with a biochemical oxygen demand value of 2.26 mg/L. The average value of biochemical oxygen demand in this Pesanggrahan River water study was 3.87 mg/L.

The highest value of the observation of chemical oxygen demand (COD) of Pesanggrahan River water is at observation point 36, which is the observation point downstream of the Pesanggrahan River flow with a COD value of 30.10 mg/L. Watershed at this point tends not to have good ecosystems with existing door conditions [5]. Pesanggrahan River water, where rubbish is carried by the Pesanggrahan River, is transported or accommodated (the existing condition at point 36 is a flood gate and a temporary "city park," kembangan) in the landfill and the lowest value of chemical oxygen demand Pesanggrahan River water is at the observation point 1 area of the observation point downstream of the Pesanggrahan River with a chemical oxygen demand value of 3.25 mg/L. This point can be said to be the cleanest point and has a tendency of Pesanggerah Watershed with a fairly good ecosystem. The average value of Chemical oxygen demand in this Pesanggrahan River water study was 14.23 mg/L. In addition, an increase in nitrate content in river water will result in an increase in chemical oxygen demand in Pesanggrahan River water. In the analysis of chemical oxygen demand values, the predictor variables of temperature and pH, there is a tendency to have no effect on the value of chemical oxygen demand, which can be seen from the results of the plot that is not spread but relatively scattered at one point (Table 1).

3.2.5 Total Dissolved Solid (TDS)

Conceptually, the amount of dissolved solids in the water will affect the light entering the water or can be said to test how much light will be inhibited when the dissolved solids are higher. The higher the value of Total Dissolved Solid, the greater the content of solids contained in water and the less light that can enter the water. This

Table 1 Values for water quality study trends for 37 representative monitoring locations from upstream to downstream of Pesanggrahan River

Sample point	Temperature	PH	TDS	TSS	BOD	COD	Hardness ($CaCO_3$)	DO	Nitrate (NO_3^-)	Nitrite (NO_2^-)	Fecal coliform	Total coliform
	°C		mg/L	mg/L	mg/L	mg/L	mg/L	mg/L	mg/L	mg/L	Cell/10 mL	Cell/10 mL
1	28.00	7.20	90	300	2.26	3.25	169.80	3.50	12.00	0.06	12.500	20.600
2	28.30	7.25	90	280	2.35	3.67	151.00	3.50	14.00	0.05	9880	22.380
3	28.70	7.40	120	192	2.58	4.80	148.20	3.40	18.00	0.03	10.000	23.700
4	28.50	7.30	120	266	2.33	5.48	184.30	3.50	16.00	0.10	9980	22.480
5	29.00	7.33	150	256	2.72	5.65	191.10	3.30	20.00	0.09	11.500	23.960
6	28.80	7.39	120	230	2.69	8.99	179.90	3.50	17.00	0.34	10.520	23.020
7	28.70	7.46	120	275	2.37	9.60	171.00	3.40	19.00	0.42	12.100	22.800
8	29.10	7.40	160	265	2.51	9.80	192.50	3.40	15.00	0.50	9600	22.100
9	29.30	7.56	100	237	2.31	9.07	214.50	3.10	15.00	0.71	12.504	20.900
10	28.80	7.50	180	165	3.02	8.42	210.80	3.10	17.00	0.80	9400	21.900
11	29.60	7.60	90	290	3.11	8.45	191.90	3.00	15.00	0.75	12.500	20.770
12	29.60	7.62	80	281	3.20	8.60	205.00	3.10	19.00	0.88	8950	19.400
13	29.80	7.61	120	274	3.00	9.21	188.40	2.90	19.00	0.79	9100	18.530
14	30.00	7.80	90	257	2.93	9.67	183.00	2.80	20.00	0.73	9350	17.600
15	29.40	7.74	180	140	3.48	10.48	190.30	2.60	25.00	0.60	9100	16.100
16	29.80	7.76	120	183	3.70	10.12	191.40	2.70	26.00	0.77	8660	16.820
17	29.40	7.50	120	182	3.51	11.84	218.10	2.80	30.00	0.95	7500	16.300
18	29.20	7.81	120	186	3.24	10.65	216.00	2.60	28.00	0.92	8110	14.600
19	28.70	7.71	110	210	3.86	11.55	207.40	2.60	28.00	1.01	7025	13.800
20	28.80	7.85	100	190	3.88	11.79	218.20	2.80	26.00	1.20	8540	12.300

(continued)

Table 1 (continued)

Sample point	Temperature	PH	TDS	TSS	BOD	COD	Hardness ($CaCO_3$)	DO	Nitrate (NO_3^-)	Nitrite (NO_2^-)	Fecal coliform	Total coliform
	°C		mg/L	mg/L	mg/L	mg/L	mg/L	mg/L	mg/L	mg/L	Cell/10 mL	Cell/10 mL
21	29.20	7.81	100	180	3.82	12.51	194.50	2.60	24.00	1.19	8330	11.000
22	28.60	7.80	100	180	4.02	12.28	209.40	2.70	31.00	1.28	8640	10.300
23	29.00	7.81	120	140	4.92	13.06	227.00	2.50	32.00	1.30	7990	6400
24	29.50	7.95	100	133	5.21	13.23	242.10	2.60	28.00	1.18	7560	8800
25	29.60	8.10	130	140	5.33	13.99	304.40	2.60	29.00	1.34	7720	9100
26	29.60	8.15	100	120	5.48	13.10	314.50	2.70	40.00	1.40	7000	8900
27	30.40	8.40	160	22	5.50	14.20	368.20	2.20	42.00	2.80	5500	8500
28	30.60	8.30	300	25	5.11	23.66	332.10	2.10	35.00	3.20	3010	5000
29	30.80	8.16	330	20	4.96	24.10	382.10	2.50	38.00	2.90	2440	4900
30	30.50	8.24	290	40	4.73	24.80	378.80	2.40	38.00	2.84	1980	4600
31	30.00	8.30	300	25	4.75	24.00	366.30	2.30	42.00	3.10	2370	4800
32	29.60	8.31	230	44	4.89	23.78	417.00	2.40	50.00	3.30	2220	4200
33	30.10	8.39	240	60	4.94	26.63	405.10	2.10	55.00	3.60	1900	3600
34	30.00	8.29	260	56	5.10	28.66	439.00	2.30	60.00	3.44	1870	3300
35	29.70	8.27	240	33	4.71	27.40	429.40	2.20	52.00	3.68	1660	2700
36	30.10	8.40	210	20	5.21	30.10	462.20	2.00	54.00	3.41	1600	2800
37	30.20	8.70	400	15	5.44	30.07	450.00	2.00	57.00	4.00	1500	2800

can result in decreased photosynthetic activity by chlorophyll organisms that are in the water [1].

The highest value of Total Dissolved Solid observation is at observation point 37, which is the observation area downstream of the Pesanggrahan River with a Total Dissolved Solid value of 400 mg/L. This observation point area is a branching point with the mookervart cengkareng/dog mogot river flow, so it can be said that Mookervart River water quality level is much lower compared to Pesanggrahan River water quality with the changes in result Total Dissolved Solid which is quite significant and the lowest value is at point 12 or at the observation point of the TPU (another name for a communal landfill) soil with a Total Dissolved Solid value of 80 mg/L. The average value of Total Dissolved Solid in this Pesanggrahan River water study was 161.89 mg/L. The high value of dissolved solids is generally caused by river areas that are close to residential areas or are in river branching areas, especially if the river has lower water quality than other river flows, therefore, that domestic waste flows into the river and increases the number of particles dissolved [10].

3.2.6 Total Suspended Solid (TSS)

The highest value from the observation of Total Suspened Solid (TSS) is at observation point 1, which is the urban upstream observation point area of DKI Jakarta Pesanggrahan River with a total Suspended Solid value of 300 mg/L, and the lowest value is at point 37 or at the observation point downstream of the Pesanggrahan River with a Total Suspended Solid value of 15 mg/L. The average value of TSS in this Pesanggrahan River water study was 159.75 mg/L.

3.2.7 Nitrate and Nitrite

The highest value of observations of nitrate content (as NO_3^-) is at observation point 33 with a nitrate value of 60.00 mg/L. The high value of nitrate at this point can be caused by the amount of domestic waste discharges of the apartments and some factory wastes around the area of the observation point, and the lowest value is at point 1 or at the urban upstream observation point in DKI Jakarta Pesanggrahan River with a nitrate value of 12.00 mg/L. The average value of TSS in this Pesanggrahan River water study was 29.89 mg/L.

The highest value of the observed nitrite (as NO_2^-) is at observation point 37, which is the observation area downstream of the Pesanggrahan River with a nitrite value of 4.00 mg/L, and the lowest value is at point 1 or at the Jakarta urban upstream observation point Sungai Pesanggrahan with a nitrite value of 0.03 mg/L. The average value of nitrite content in the Pesanggrahan River water study was 1.50 mg/L.

The low content of nitrates and nitrites in the surface layer is because in the surface layer of the river the available oxygen is quite abundant in the presence of oxygen diffusion from the atmosphere. With the help of bacteria, the oxygen will oxidize nitrites into nitrates; therefore, the nitrite content in the nitrite layer becomes nitrate and the nitrite content in the surface layer becomes small. Thus, the value of the nitrite content is inversely proportional to the value of the oxygen content in the water. This can be proven by reviewing the value of dissolved oxygen (DO) and also the value of the fecal content and Total Coliform which are interrelated.

3.2.8 Hardness

Hardness is derived from salt chemical compounds in the form of cations and anions, causing the water quality of the Pesanggrahan River to become turbid and have a density when viewed with the naked eye.

The highest value of the observation of the Pesanggrahan River water hardness is at observation point 36 which is the observation point of the downstream floodgate Pesanggrahan River (Kembangan) with a water hardness value of 462.20 mg/L which could be due to the floodgates holding up the flow of the Pesanggrahan River so that sediment occurs at the bottom of the river, which causes the depth of the existing deposits [2]. Deposits that have a lower density than water usually arise to the surface, which causes the physical condition of the water to become turbid and hardness in the water to appear, and the lowest value is at point 3 or at a location near the downstream of the Pesanggrahan River with a hardness water value of 148.20 mg/L, and this condition occurs due to the concentration and activity of domestic and industrial waste production which is still minimal, but supported by low water flow relatively swift compared. The highest value of the observation of the Pesanggrahan River water hardness is at observation point 36 which is the observation point of the downstream floodgate Pesanggrahan River (Kembangan) with a water hardness value of 462.20 mg/L which could be due to the floodgates holding up the flow of the Pesanggrahan River so that sediment occurs at the bottom of the river, which causes the depth of the existing deposits. Deposits that have a lower density than water usually arise to the surface, which causes the physical condition of the water to become turbid and hardness in the water to appear, and the lowest value is at point 3 or at a location near the downstream of the Pesanggrahan River with a hardness water value of 148.20 mg/L, and this condition occurs because this condition is due to the concentration and activity of domestic and industrial waste production which is still minimal, but supported by low water flow relatively swift compared the average value of hardness content in the Pesanggrahan River water study was 263.38 mg/L. This shows the Pesanggrahan River water flow factor which has a relatively low speed, which is evidenced by the relatively high hardness level of the Pesanggrahan River and the flow of the Pesanggrahan River does not carry over [11].

3.2.9 Fecal Coliform and Total Coliform

The highest value of Fecal Coliform observations of Escherichia Coli bacteria or often called E-coli is at observation point 9, which is the RPTRA Anggrek Bintaro-observation point with a Fecal Coliform value of 12,504 cells/100 mL (this area is

Fig. 1 (Domestic waste) Water canal of an apartment building in the Pesanggrahan River

a densely populated residential/residential area) and the lowest value is located at point 37 or downstream of the Pesanggrahan River with a Fecal Coliform value of 1500 cells/100 mL. The average value of Fecal Coliform in the Pesanggrahan River water study is 7300 cells/100 mL (Figs. 1 and 2).

The highest value of bacterial Total Coliform observations was at observation point 5, namely the observation point of Pondok Pinang toll gate area with a Fecal Coliform value of 23,960 cells/100 mL (this area has a dense urban forest/green land with a deep cliff wall that is deep enough, so still there are many natural ecosystems in it), and the lowest value is at points 36 and 37 or downstream of the Pesanggrahan River with a Fecal Coliform value of 2800 cells/100 mL. The average value of Fecal Coliform in the Pesanggrahan River water study is 13,290 cells/100 mL. The results of the review of the existing field conditions, the value of bacterial coliform which is above the average value is a residential/residential area wherein the area of domestic waste comes from organic sewage wastes, which is known if bacteria such as E-Coli can multiply rapidly or sourced from the dung of living things (organic).

Decrease in the value of Fecal Coliform is directly proportional to the decrease in the value of Total Coliform, and inversely proportional to the value of Pesanggrahan River water hardness. Drastic decline in organisms such as bacteria/coliforms is generally caused by the content of salt chemical compounds that are sourced from the production of industrial wastes and also detergent/domestic waste, which are the constituent compounds.

Fig. 2 (Industrial waste) home textile industry waste (washing and staining of jeans)-Black solid color flowing into pesanggrahan river water flow

3.3 *Factor of Declining Pesanggrahan River Water Quality*

The decline in the water quality of the Pesanggrahan River is due to the large number of violations committed by the community, especially many who use residential buildings that have become business locations that produce domestic wastes discharged into the Pesanggrahan River. As an example of the Kebayoran Lama and Pesanggrahan area, houses around the Pesanggrahan Watershed are used as the location of the textile industry where the disposed wastes are directly channeled into the Pesanggrahan River, although in their designation and licensing they are both are the same. Considering the can be considered as industrial land that is illegal and violates spatial planning. Decreasing water quality in the Pesanggrahan River has accumulated chemical contents so that the value of water quality will get worse in the downstream of the river. The decline in the water quality of the Pesanggrahan River caused in large part by domestic waste which is die to the poor sanitation and drainage system in the Pesanggrahan River Region where houses dispose their domestic wastes directly into drainage connected to the Pesanggrahan River without any process of processing waste either independently or communally the water does not meet the applicable quality standards.

The wastewater disposal system is in the form of domestic (household) wastewater, which domestic wastewater management is generally carried out by using a local sanitation system (on site sanitation) in the form of latrines, which are managed individually or communally, which are equipped with septic or cubluk tanks [7]. Meanwhile, if using a centralized sanitation system (offsite sanitation), the implementation requires a considerable amount of cost also in its maintenance, so that the use of the existing centralized sanitation system is estimated to only be able to serve residents living in the DKI Jakarta city [8].

4 Conclusion

The water quality of the Pesanggrahan River tends to be low and polluted. This can be seen from the observations of the colors of the Pesanggrahan River water samples compared to Aquadest or distilled water, all of which were found at the point of observation for deposits and discoloration. From the water quality index value, chemical and biological contents in the Pesanggrahan River water show that the Pesanggrahan River water is included in the polluted river category. The level of water pollution in the Pesanggrahan River is greater in the downstream river due to the accumulation which is carried by the river water flow of chemical compounds due to industrial and domestic waste from the upstream of the river. Most of the Watershed Houses are both the same. Spatial allotment errors occur mostly in industrial activities in residential areas. The development that occurred in the Pesanggrahan Watershed area was more on the increase in the number of house buildings and the condition of vacant land to become vertical housing and commercial buildings. The development occurred almost entirely in the period of 2004–2010, while in the period 2010–2013, more changes were made in the structure of the Pesanggrahan River flow, namely the widening and straightening of the river.

References

1. American Public Health Association. (1999). *Standard methods for the examination of water and wastewater* (20th ed.). Baltimore: American Public Health Association.
2. Campbell, J., & Peterson, D. (2010). Determination of water hardness from common water sources using flame atomic absorbance spectrometry. *Concordia College Journal of Analytical Chemistry, 1*, 4–8.
3. Griffiths, P. G., Hereford, R., & Webb, R. H. (2006). *Sediment yield and runoff frequency of small drainage basins in the Mojave Desert.* California and Nevada: USGS Fact Sheet: Recoverability and Vulnerability of Desert Ecosystems. Retrieved from https://www.fondriest.com/environmental-measurements/parameters/water-quality/water-temperature/
4. Kannel, P., Lee, S., Lee, Y.-S., Kanel, S., & Khan, S. (2007). Application of water quality indices and dissolved oxygen as indicators for river water classification and urban impact assessment. *Environmental Monitoring and Assessment, 132*, 93–110.
5. Lumaela, A., Otok, B., & Sutikno. (2013). Pemodelan chemical oxygen demand (COD) Sungai di Surabaya Dengan Metode Mixed Geographically Weighted Regression. *Jurnal Sains Dan Seni Pomits, 2*, 100–105.
6. Maryono, A. (2017). *Water resources management reform.* Yogyakarta: Gadjah Mada University Press.
7. Maryono, A. (2018). *River border area management with integral approach: Regulation, institutional, spatial, social, morphological, ecological, hydrological and engineering*. Yogyakarta: Gadjah mada University Press.
8. Nazir, M. (2009). *Metode Penelitian.* Jakarta: Perpustakaan Nasional.
9. Odum, E. (1971). *Fundamental of ecology*. Philadelphia: W.B. Saunder Com.
10. Scannell, P. W., & Jacobs, L. L. (2001). *Technical report: No. 01-06 effects of total dissolved solids on aquatic organisms.* Alaska: Alaska Department of Fish and Game: Division of Habitat and Restoration.

11. Wood, M. S. (2014). *Estimating suspended sediment in rivers using acoustic Doppler meters*. Retrieved Mei 25, 2019, from U.S. Geological Survey Fact Sheet: https://www.fondriest.com/environmental-measurements/parameters/water-quality/turbidity-total-suspended-solids-water-clarity/.

Spatial Assessment of Micro-hydropower Plant in Subsidized Housing Case: Perumnas Parung Panjang

Andrean Eka Lucianto and Herdis Herdiansyah

Abstract Indonesia is a tropical country with rainfall, tropical forest, springs, and rivers that flow, which give this country a potential to generate an electricity and use micro-hydropower plant (MHPP) to subsidized housing, especially in Perumnas Parung Panjang. This research applies to multi-method research. Primary data collection was performed through interview to determine the social and economic conditions of the community related to purchasing power, level of education, and public interest in MHPP. That data is carried out by an in-depth interview method. However, collection of secondary data in the form of land physical data (topography, soil type, land cover) and rainfall data was obtained from relevant government agencies. The result was obtained using spatial scoring and SWOT analysis. The results of this study are the potential of Parung Panjang Housing to apply the concept of MHPP from the physical aspect of the area, supported by the existence of two rivers that pass through the housing, approved with 88 points of spatial scoring. However, from the social aspect, the community is still declined to apply the micro-hydropower plants, due to the lack of public knowledge about the technology. In addition, from the economic aspect of the community, which is in the middle–lower income, an investment is considered too burdensome. With various benefits that can be obtained, if the concept of MHPP is implemented, the solution is that the government assists an education and subsidies related to MHPP development in subsidized housing.

Keywords Micro Hydro Power Plant · Subsidized housing · Spatial scoring

1 Introduction

Indonesia still has a problem in using renewable energy resources, which 96% energy comes from fossil fuel (48% fuel, 30% coal, and 18% gas) and only about 4% is energy from renewable energy [1]. In terms of the geographical position of Indonesia, it is located in equatorial zone, which means this tropical country is hot and humid, so it

---

A. E. Lucianto · H. Herdiansyah (✉)
School of Environmental Science, Universitas Indonesia, Jakarta 10430, Indonesia
e-mail: herdis@ui.ac.id

S. S. Emamian et al. (eds.), *ICSDEMS 2019*,
https://doi.org/10.1007/978-981-15-3765-3_13

can rain throughout the year, and it has some rivers, lakes, and tropical forests that keep the water resources in the ground. Due to that natural potency of the great water resources in Indonesia, micro-hydropower plant can be developed widely around the country and become an alternative solution for energy supply [2–5]. So, Indonesia has great potential in hydro power plant (75.000 MW), but the development of the hydro power plant is still very small (9% from the total potential) [1]. Besides that, the biggest consumption of electricity was residential with 93.6 MW or 43.3% in 2016 [6]. So, this research focuses on knowing the potential of using micro-hydropower plant in the housing, especially in subsidize housing.

Perumnas Parung Panjang is a subsidized housing. However, the middle and lower classes are settled in this area, who must have a good level of resilience in their electricity needs and increasing community assets by using a micro-hydropower plant in their area [7]. Micro-hydropower plant is the most beneficial of renewable energy power plants, because some real steps can be implemented to solve some obstacles in the development process, so it will match with the needs of subsidized housing, which is needed to be affordable [1]. This research showed the suitability of physical regional aspect to support the MHPP and the social-economy aspect of people who settled in this area and gives some suggestions, if the private developers, NGO or government, want to develop MHPP in there. Then, the application of MHPP as a source of electricity will help the middle–low residents to increase their economic level in the future.

2 Methods

The study area, especially Perumnas Parung Panjang Housing, is located in Tangerang, West Java, Indonesia. The area is 214.0 ha and has around 6.092 households. There is one big river that branches into two rivers across the area. Study area location is shown in Fig. 1.

Two data types of this research are physical regional data and socio-economy regional data. The physical regional data is spatial data that contains oil type, rainfall level, topographic contour, and land cover data. The physical data collected from government authorities' agency, which provide the data legally, is shown in Table 1. The spatial data is analyzed by geographic information system (GIS) from ArcMap 10.2.

The methodology for the assessment of potential application of MHPP in Perumnas Parung Panjang is presented in Fig. 2. This method is the first method to measure the potential of using MHPP that combines physical land aspect and social-economy aspect. Some researchers have studied the physical land data only [8], and the others have used the descriptive of data [1]. This research used a quantitative approach, which applies multi-method research. Primary data collection through an in-depth interview was used to determine the social and economic conditions of the community related to purchasing power, level of education, and public interest in MHPP. The secondary data in the form of land physical data (topography, soil type,

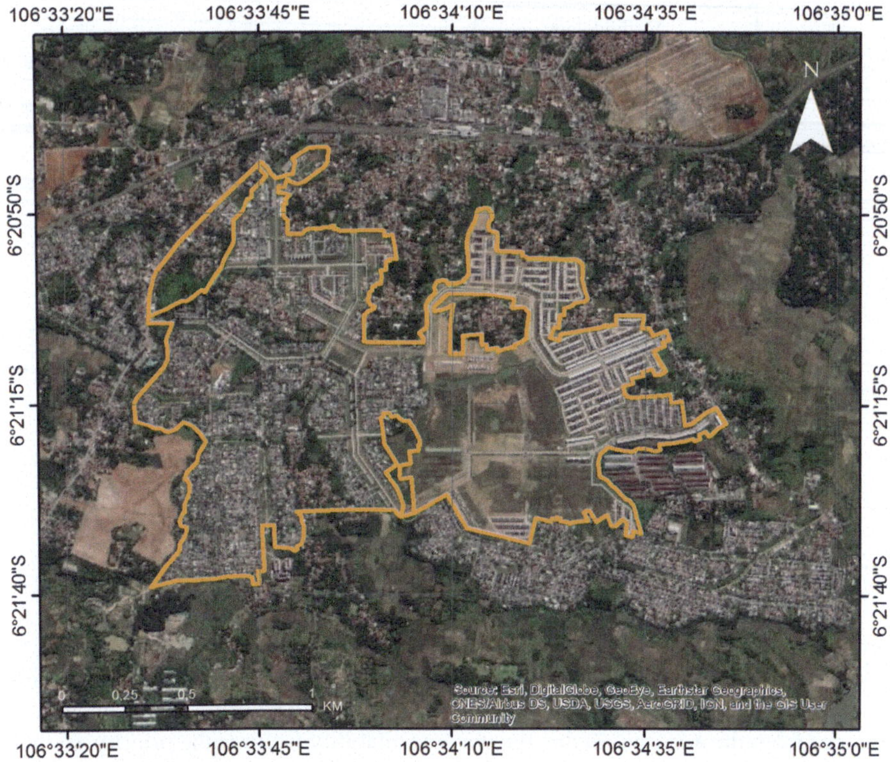

Fig. 1 Study area in Perumnas Parung Panjang

Table 1 Data and source

Data	Source
Soil type	Bogor Regional Planning Agency
Rainfall level	Bogor Regional Planning Agency or Meteorology and Geophysics Climatology Agency
Topographic contour	Agency for Geospatial Information
Land cover	Bogor Regional Planning Agency

and land cover) and rainfall data was obtained from relevant government agencies. The data was analyzed using scoring analysis. Finally, the researcher used SWOT analysis to get the result and some development advices.

The physical aspect of the area is supported by the existence of two large rivers that pass through the housing. Therefore, to measure the potential of that river, a study from the water catchment area is needed. There are three parameters that are the input of the study including the soil data, rain level data, and topography data. The parameters have a score that compares the level of the effect it has on water

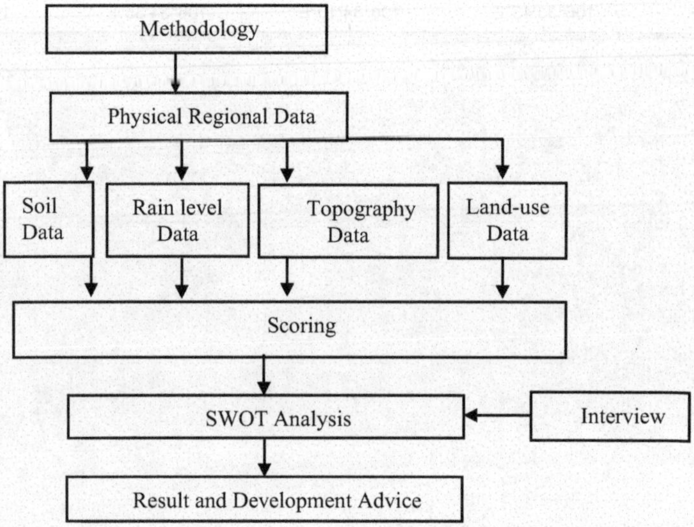

Fig. 2 Flowchart of the methodology

supply in the river (Table 2). The score of each parameter is calculated by Eq. 1, in which STSP is subtotal score of parameter, S_p is score of parameter, S_i is score of indicator, and % Area is area size of each indicator. The score of parameters is obtained from the assumption that the water availability factor is the priority, so that rainfall parameter is placed in the most affected and has the score of 40, while the other parameters affect in terms of speed permeated the water into the ground. The soil parameter has the score of 20, topography parameter has the score of 20, and

Table 2 Scoring for the parameters and indicators

Score of parameters	Parameters	Score of indicators	Indicators
40	Rainfall	100	>1000 mm/years
		75	500–1000 mm/years
		50	<500 mm/years
20	Soil	100	Podzolik soil
		75	Andosol soil
		50	Grumosol soil
20	Topography	100	Slope > 30°
		75	Slope 15–30
		50	Slope < 15°
20	Land cover	100	Housing, industry
		75	Grassland
		50	Agriculture, forest

Table 3 Physical regional total score

Score of parameters	Parameters	Score of indicators	Indicators	% Area of indicator	Score
40	Rain level	100	>1000 mm/years	100	40
		75	500–1000 mm/years	0	0
		50	<500 mm/years	0	0
20	Soil	100	Podzolik soil	100	20
		75	Andosol soil	0	0
		50	Grumosol soil	0	0
20	Topography	100	Slope > 30°	0	0
		75	Slope 15–30	20	3
		50	Slope < 15°	80	8
20	Land cover	100	Housing, industry	60	12
		75	Grassland	20	3
		50	Agriculture, forest	20	2
Total score					88

land cover parameter has the score of 20. Then, the score of indicators is obtained from the same assumption with the parameters.

$$STSP = Sp \times Si \times \%Area \tag{1}$$

3 Result and Discussion

3.1 Rain-Level Score

Perumnas Parung Panjang is located in Bogor Region that is called "The City of Rain" from local people, and this name matches with the data that all of area has rain level above 2500 mm/year, even in some areas rain level is of 5000 mm/year. Data is obtained from Bogor Regency Spatial Planning 2012/*Rancangan Tata Ruang Wilayah Kabupaten Bogor tahun 2012–2032*, especially for the rain level. The score of Perumnas Parung Panjang for rain level is presented in Table 3.

3.2 Soil Data Score

Bogor Region is near a volcanic mountain, which makes this area has soil that rich of minerals. Three types of soil are dominant: andosol soil, podzolic soil, and grumosol soil. Perumnas Parung Panjang has podzolic soil. Data is obtained from Bogor Regency Spatial Planning 2012/*Rancangan Tata Ruang Wilayah Kabupaten Bogor tahun 2012–2032.* [9], especially for soil data. The score of Perumnas Parung Panjang for soil is presented in Table 3.

3.3 Tophography Score

Bogor region is surrounded by mountains and hills along the south. There are steep slopes in the south and tend to be sloping in the north. Perumnas Parung Panjang is located in the north side of Bogor region, so that area tends to be flat generally, but has differences in the height of the land which is quite bumpy in some places. Data is obtained from internal analytical research of Agency for Geospatial Information/Badan Informasi Geospasial. The score of Perumnas Parung Panjang for topography is presented in Table 3.

3.4 Land-Use Score

Land use in Perumnas Parung Panjang is a reflection of Bogor Regional Planning that contained in Bogor Regency Spatial Planning 2012/*Rancangan Tata Ruang Wilayah Kabupaten Bogor Tahun 2012–2032* [9]. The research area has several section areas such as a settlement area, green open space, watershed, park, school, road, and place of worship. The score of Perumnas Parung Panjang for land use is presented in Table 3.

3.5 Physical Regional Total Score

Physical regional aspect of Perumnas Parung Panjang contains the rain-level data, soil data, topography data, and land-use data. The calculation of the physical regional total score is presented in Table 3, based on data and description in Sects. 3.1–3.4. The calculation of % area of indicator used ArcMap 10.2 to measure the area (m^2), then it was compared to the total area to get the percentage. The result of physical regional total score is 88 based on Table 3. The physical aspect of the region is very suitable, based on Table 4.

3.6 SWOT Analysis

SWOT analysis of the potential implementation of MHPP (Table 5).

Table 4 Result score classification of physical aspect

Score classification of physical aspect	Range score
Very bad area	0–25
Bad area	26–50
Suitable area	51–75
Very suitable area	76–100

Table 5 SWOT analysis

	Strength The spatial analysis delivers that regional aspect of Perumnas Parung Panjang has potential of development of MHPP	*Weakness* Middle–low income society that inhabit in Perumnas Parung Panjang has limitation of economy for fund and has less technology knowledge, especially MHPP
Opportunity Perumnas Parung Panjang is building a new cluster that collaborates between Perum Perumnas (state-owned company) and BSA Land Property (private company)	The development of MHPP in Parung Panjang, can be created with collaboration of Perumnas as the main developer in this area and BSA Land as private a developer	The middle–low society must be subsidized from private company or government
Treat The bad habit from some inhabiters in the society that throw garbage into the river	The developer or NGO give some counseling/socialization to the society to keep the river clean	The infrastructure of waste processing must be developed to anticipate the habits, and the river stays clean to be the water supply for MHPP

The steps that can be taken from analysis SWOT:

1. The step from the strength is that the government/NGO needs to give some counseling/socialization about the potential of MHPP to settlements in the Perumnas Parung Panjang. Communities need to be aware that the existence of two rivers in their area has physical potential that supports MHPP utilization in their area [5].
2. The step to minimize the weakness is that the government/NGO needs to provide subsidies in MHPP development in Perumnas Parung Panjang.
3. The step to maximize the opportunity is that the main developer (Perum Perumnas) should create an agreement with the second developer (BSA Land Property) in order to make commitment to implement the MHPP in Perumnas Parung Panjang.
4. The step to reduce impact from the threat is that the government/NGO needs to give some counseling/socialization about the importance of protecting and keeping the environment for settlements in the Perumnas Parung Panjang, so that the communities will participate a role in reducing the impact of climate change.

4 Conclusions

Perumnas Parung Panjang has a potential to implement MHPP from the physical area aspect, with 88 points of spatial score. But, there is a resistance from the social-economy aspect of the residents. The limitation of economy from the middle–low income residents has implications to the level of technology knowledge and ability to pay for the fund. The government/NGO supposed to give counseling to the residents about the benefits. Besides that, the government needs to give subsidies that are related to MHPP in Perumnas Parung Panjang.

Hopefully, there is the NGO/government that has a willingness to create some counseling and to build the MHPP in the future. It absolutely can help the middle–low income residents to be more independent in their energy needs.

Acknowledgements This research is funded by Decentralization Research—Ministry of Research, Technology and Higher Education of the Republic of Indonesia—Universitas Indonesia (Riset Desentralisasi—Kementrian Riset Teknologi dan Pendidikan Tinggi—Universitas Indonesia) [contract number: NKB-1903/UN2.R3.1/HKP.05.00/2019].

References

1. Erinofiardi, A., Gokhale, P., Date, A., Akbarzadeh, A., Bismantolo, P., Suryono, A. F., Mainil, A. K., Nuramal, A. (2016). A review on micro hydropower in Indonesia. *Energy Procedia, 110*(316–321). https://doi.org/10.1016/j.egypro.2017.03.146
2. Sriyana, I. (2018). Evaluation of micro hydro power plants in Central Java toward sustainability against hydrology condition of watershed. In *E3S Web of Conferences* 73, 01017. https://doi.org/10.1051/e3sconf/20187301017
3. Suprapty, B., Malani, R., Minardi, J. (2018). Rainfall prediction using fuzzy inference system for preliminary micro-hydro power plant planning. *IOP publisher: Earth and environmental Science* 144. https://doi.org/10.1088/1755-1315/144/1/012005
4. Yuhendra, Zaini. (2018). GIS-based mapping potential sites for micro-hydro power plants in West Sumatera. *2018 IEEE International Conference on Innovative Research and Development (ICIRD)*, pp. 1–4. Bangkok, 2018. https://doi.org/10.1109/ICIRD.2018.8376301
5. Satria, H., Naldo, N., Krypton, A., Adanta, D; Budiarso, B. (2018). Diffusing signage as a visual communication in applied of Pico Hydro technology in Batu Roto, Bengkulu Province, Indonesia. *ASEAN Journal of Community Engagement*, 2(1). https://doi.org/10.7454/ajce.v2i1.108
6. Dirjen. Ketenagalistrikan ESDM. (2017). *Statistik Ketenagalistikan 2016*. Jakarta: Kementrian ESDM
7. Gippner, O., Dhakal, S., Sovacool, B. (2013). Microhydro electrification and climate change adaptation in Nepal: socioeconomic lessons from the rural energy. *Journal Mitigation and Adaptation Strategies for Global Change 04/2013, 18.* :https://doi.org/10.1007/s11027-012-9367-5
8. Ratnata, I., Saputra, W., Somantri, M., Mulyana, E., Ardhika, A. (2017). Preliminary study of micro-hydro power plant (MHPP) in the rural area. In: IOP Conference Series: Material Science and Engineering, 384. https://doi.org/10.1088/1757-899x/384/1/01/2067
9. Regional Planning Agency of Bogor Region. (2016). Peraturan Daerah Nomor 11 tahun 2016 tentang RTRW Kabupaten Bogor. Bappeda Bogor.

Cost-Benefit Analysis of Combined Retaining Walls Construction

Mohd. Sukry Mohamed and Samira Albati Kamaruddin

Abstract Stability of slopes can be improved by using soil nailing as one of in situ soil reinforcements. In the case when the soil nailing system alone is not feasible, then other types of retaining walls can be used together. Therefore, this paper aimed to assess combination soil nailing and crib wall in terms of cost benefits. Analysis of cost comparison from cost-benefit analysis (CBA) was carried out between soil nailing and crib wall as a reference. Common problems of soil nailing such as rebar encroaching to neighbouring land were solved by using a crib wall because it can be constructed on a small space. This type of combination is known as a combined system retaining structure. Design and construction data have been collected from a project under construction in Selangor. The results of CBA proved that the construction should proceed with the combined system when the benefit-cost ratio (BCR) is greater than 1 and the net present value (NPV) is greater than 0. When NPV and BCR are used to determine the CBA, there is an economic validation to further the project if NPV and BCR are equal to 0 that shows the breakeven point (less certain outcome). However, if BCR is less than 0, then the project should not proceed with the selected method. The findings suggest that a comparison of the cost estimation could verify the future feasibility of the proposed method.

Keywords Soil nailing · Crib wall · Cost-Benefit analysis · Cost saving · Cost-effective

1 Introduction

Soil nailing system has been used for strengthening soil for slope, excavation, and retaining wall [1–9]. However, in some cases, the soil nailing system alone may not be applicable and may be considered costly to be used as the overall retaining wall system. If the nail encroaches to the neighbouring land, it will trespass to the boundary and create unlawful intrusion. The most economical solution to this problem can

Mohd. S. Mohamed (✉) · S. A. Kamaruddin
Department of Engineering, Razak Faculty of Technology and Informatics, Universiti Teknologi Malaysia Kuala Lumpur, Kuala Lumpur, Malaysia
e-mail: sukry.maey@gmail.com

S. S. Emamian et al. (eds.), *ICSDEMS 2019*,
https://doi.org/10.1007/978-981-15-3765-3_14

be solved by combining the soil nailing and crib wall as a combined system. The retaining wall system that combines two or more retaining structure system will bring more benefits to the project [10, 11]. The first step in common slope stabilization work is to determine whether retaining wall can be constructed in the cut or fill slopes [11]. For a retaining wall system, crib wall can be constructed in fill slope and soil nailing which are more suitable in cut slope. By selecting a combined retaining wall, advantages and efficient results bring substantial cost savings and reduced construction period. Therefore, the construction cost of slope stabilization work can be compared among soil nailing and other options of retaining walls such as crib wall. In evaluating the comparison, cost-benefit analysis (CBA) can be used to determine whether both retaining walls can bring benefits or losses as slope stabilization options in the project. The CBA is a well-developed tool for supporting decision-making processes [12].

2 Methodology

The CBA offers an analysis of the strengths and weaknesses evaluation of a slope stabilization project. The overall results of CBA can be determined using two methods [13]. The first one is the net present value (NPV) and the second one is the benefit-cost ratio (BCR). Both methods are used as an indicator of CBA [14]. All benefits and costs can contribute to a monetary value and can be significantly shown by the CBA that allows comparison. The NPV of an undertaking is equivalent to the distinction between the present estimation of the advantages, Eq. (1) and the present estimation of the expenses, Eq. (2). The project can be continued if the benefit exceeds the cost from economic justification. If the value of BCR is less than 1.0, it would not be recommended to proceed with the construction method. The higher the benefits associated with the considered alternative, the bigger the value and it will be more than 1.0.

The highest BCR would be preferred in project construction. When NPV and BCR are used to determine the CBA, there is an economic validation to further the project if NPV and BCR are equal to 0 that shows the breakeven point (less certain outcome). However, if BCR is less than 0, then the project should not proceed with the selected method. As shown in Eq. (3), the NPV is calculated by subtracting the sum of the costs with the sum of the benefits. The BCR in Eq. (4) provides a ratio value for the CBA by dividing the sum of the total benefits by the sum of the total costs.

$$\text{Present value of Benefits, } \mathrm{PV}(B) = \text{Time-saving} + \text{Reduce injury/Lifesaving} + \text{Create job} \tag{1}$$

$$\text{Present value of Costs, } \mathrm{PV}(C) = \text{Consultant design fees}$$

Table 1 Comparison of machinery cost, materials, construction of soil nailing and crib wall according to a common bill of quantity in Malaysia

Specification of construction	Soil nailing	Crib wall
Machinery and transportation	Plant and equipment	Mobilisation of the crib wall
Construction materials	Soil nails, nail head, shotcrete	Precast concrete components, aggregates/granular
Supply and installation	Horizontal drain, berm drain	Crib wall installation, RC levelling pad/piled slab preparation and casting, granular backfilling
Testing	Pull out test	Mackintosh test

$$+ \text{Material and Construction} + \text{Maintenance} \quad (2)$$

$$\text{NPV} = \Sigma\text{PV}(B) - \Sigma\text{PV}(C) \quad (3)$$

$$\text{BCR} = \Sigma\text{PV}(B) \div \Sigma\text{PV}(C) \quad (4)$$

Soil nailing and crib wall are easy to construct, require fewer skilled workers, use no heavy machinery and little maintenance if planted scrubs for aesthetic [15, 16]. Therefore, the cost of machinery, materials and manpower are more economical compared to the other type of retaining walls. The comparison of these construction costs is based on the specifications in Table 1.

3 Results and Discussion

Cost of both systems was calculated to directly make a comparison of the decision implication. In the comparison of the installation cost of soil nailing, the calculation was estimated for a 100 m length of needed materials. All of the rating data was taken from the bill of quantity provided for a slope stabilization project in Selangor. However, the installation cost of the crib wall considered the rate of construction taken from another slope project in the same state. To be fair, all the estimated construction cost is calculated based on a 100 m length of retaining wall system. Construction cost consists of the cost of machinery, materials and manpower. From the comparison, it is proved that the cost of construction soil nailing system is much lower than the crib wall system. The cost difference is RM 329,057 and it is 33.35%. Constructing soil nailing system is supposed to be cheaper than the crib wall system, but purchasing a piece of land would increase the overall cost. The results in Fig. 1 show that the construction cost of soil nailing is only RM 657,578, while the construction cost of the crib wall is RM 986,635. However, when the land purchase is added, the total cost of RM 5,381,960 is increased by 88%. As can be seen, the CBA indicated that the project should proceed and the combined system is more preferable [17]. The combination

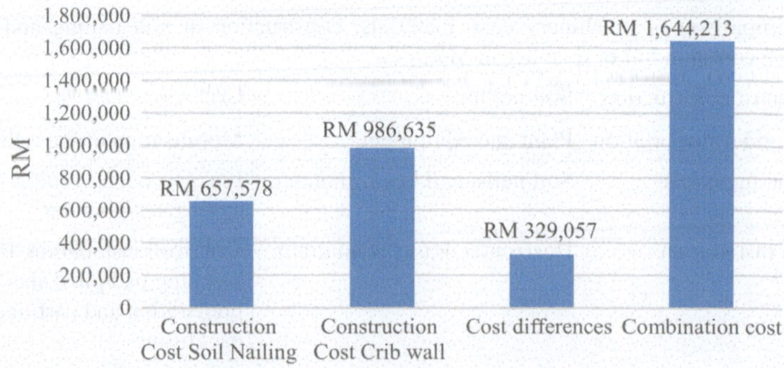

Fig. 1 Comparison between construction cost of soil nailing and crib wall

Table 2 Comparison of costs building for 100 m length of retaining walls and 2 years maintenance after project completion

The requirement for cost and benefits	Soil nailing	Crib wall
Consultant fees for design	RM 39,455	RM 59,198
Land purchase	RM 5,381,960	N/A
Materials and construction	RM 657,587	RM 986,635
Maintenance landscape	RM 9600	RM 3200
Time of saving	2 months	1 month
Saving money	RM 16,214	RM 12,164
Lifesaving or reduced injury	RM 184,900	RM 438,000
Pollution reduction	RM 1,460,000	RM 1,460,000
The initial duration of job creation	2 years	4 months
Labour cost	RM 184,900	RM 72,000
Present value of benefits, PV (*B*)	1,846,014	1,982,164
Present value of cost, PV (*C*)	6,078,993	1,049,033
Net present value, NPV	−4,232,979	933,131
Benefit-cost ratio, BCR	0.30	1.89

cost by both system results of RM 1,644,213 shows that soil nailing construction with the land purchase is not cost-effective. Thus, a combined system as an effective solution should be implemented for the slope strengthening and stabilization system.

Table 2 shows the costs and benefits of soil nailing and crib wall. The design cost of 6% is used as stated in the Board of Engineers Malaysia (BEM) scale of fees (Part B). The land price is taken from the current rate of area in this project per meter square. The 100 m length of soil nailing wall with 10 m width is constructed with 9 m nails length in addition to 1 m clearance. The construction cost is calculated based on the bill of quantity rate, maintenance for the newly purchased land and rectification of construction defects. Daily basis labour work is applied in the calculation. Duration

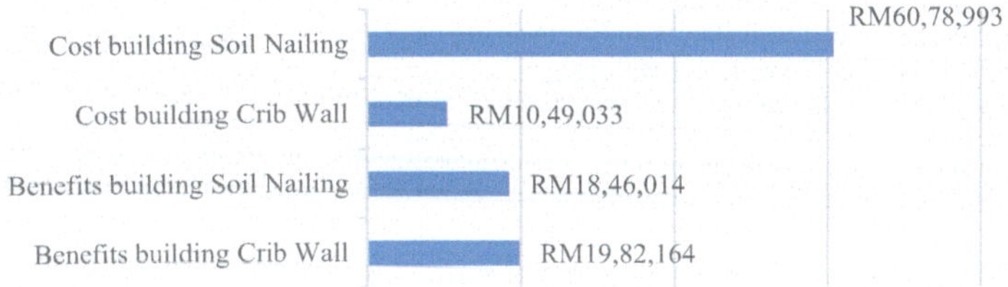

Fig. 2 Building cost and benefits of soil nailing and crib wall

of the project is assumed to be 120 days, the defects liability period, maintenance and monitoring are 2 years after project completion.

Figure 2 shows a cost comparison between soil nailing and crib wall. The costs to build soil nailing and crib wall system are RM 6,078,993 and RM 1,049,033, respectively. The cost of building the retaining wall consists of design fees, land purchase, construction, maintenance landscape and rectification defects. Cost building of retaining wall and PV (C) gives the same indicator. PV (B) is the benefits of building a retaining wall and consists of time-saving, reduce injury/lifesaving, pollution reduction and job creation. The calculated cost difference is RM 5,029,960 or 82.74%. The cost of building soil nailing system increased more than eight times from the total overall cost due to neighbouring land purchase. Therefore, the benefit of building soil nailing system of RM 1,846,014 is lower than the benefit of building crib wall of RM 1,982,164. The cost difference between the two options is RM 136,150 or 6.87%.

4 Conclusions

Soil nailing is a fast slope strengthening method and cost-effective. However, there are some disadvantages to performing this method. It is difficult to solve nail encouraging to neighboring land. In this case, the recommended crib wall can offer a better solution as it only requires small space of construction. The combination system that is known as a combined system has become a selective solution to the industry. The crib wall system is proposed if the client encounter limits the land spaces and boundary issues. The findings suggest that the crib wall system should be selected because the NPV and BCR result in more than 1.0. If the soil nailing system is solely used, an additional cost will be needed to purchase land. An extra budget is also needed to maintain the new land. Some new land purchases are very expensive due to the higher land value in the urban area. On the other hand, no additional land is required if the combined soil nailing and crib wall system can be built together. Therefore, a combination soil nailing and cribwall as a retaining wall system can completely accommodating each other weaknesses and become a better solution to stabilize slope and offer a range of engineered solutions to the clients.

Acknowledgements The first author is an Engineering Doctorate student who is partially supported by the Ministry of Higher Education (Malaysia).

References

1. Fan, C. C., & Luo, J. H. (2008). Numerical study on the optimum layout of soil–nailed slopes. *Computers and Geotechnics, 35*(4), 585–599.
2. Giri, D., & Sengupta, A. (2009). Dynamic behavior of small scale nailed soil slopes. *Geotechnical and Geological Engineering, 27*(6), 687.
3. Yu, Y. X. (2010). *Study of the working characteristics in deep foundation pit supported by composite soil-nailed wall (Order No. 10341386).* Available from ProQuest Dissertations & Theses Global (1870503681).
4. Huang, J. H. (2012). Finite element analysis of composite soil-nail retaining structure in foundation pit engineering. In W. J. Yang & Q. S. Li (Eds.), *Progress in Industrial and Civil Engineering, Pts. 1–5, 204–208,* 163–167.
5. Giacchetti, G., Grimod, A., & Cheer, D. (2013). Soil nailing with flexible structural facing: design and experiences. In C. Margottini, P. Canuti, & K. Sassa (Eds.), *Landslide science and practice* (Vol. 6, pp. 655–660)., Risk assessment, management and mitigation Berlin: Springer.
6. Kim, Y., Lee, S., Jeong, S., et al. (2013). The effect of pressure-grouted soil nails on the stability of weathered soil slopes. *Computers and Geotechnics, 49,* 253–263.
7. Ghareh, S. (2015). Parametric assessment of soil-nailing retaining structures in cohesive and cohesionless soils. *Measurement, 73,* 341–351.
8. Lin, P., & Liu, J. (2017). Analysis of resistance factors for LFRD of soil nail walls against external stability failures. *Acta Geotechnica, 12*(1), 157–169.
9. Seo, H., Lee, I.-M., Ryu, Y.-M., et al. (2019). Mechanical behavior of hybrid soil nail-anchor system. *KSCE Journal of Civil Engineering.* https://doi.org/10.1007/s12205-019-2268-3.
10. Department of Transportation Federal Highway Administration. (2015). *Soil nail walls reference manual.* U.S. Publication No. FHWA-NHI-14–007, FHWA GEC 007.
11. Eldiasty, W. A., Altahrany, A. I., & Elmeligy, et al. (2019). Comparison between monotype and hybrid earth retaining structures. *Innovative Infrastructure Solutions, 4*(1), 30. https://doi.org/10.1007/s41062-019-0213-4.
12. Belay, A. M., Torp, O., Thodesen, C., et al. (2016). A framework for organizing a resilient cost benefit analysis for construction projects. *Procedia Engineering, 145,* 1169–1176. https://doi.org/10.1016/j.proeng.2016.04.151.
13. National Highway Institute, Md. Lani, N. H., Syafiuddin, A., & Yusop, Z. (2019). *Comparison of cost benefits of new installation and retrofitted rainwater harvesting systems for commercial buildings.* Cham.
14. Singh, B. B., Kostoulas, P., Gill, J. P. S., et al. (2018). Cost-benefit analysis of intervention policies for prevention and control of brucellosis in India. *Plos Neglected Tropical Diseases, 12*(5), 16. https://doi.org/10.1371/journal.pntd.0006488.
15. Acharya, M. S. (2018). Analytical approach to design vegetative crib walls. *Geotechnical and Geological Engineering, 36*(1), 483–496.
16. Rabie, M. (2016). Performance of hybrid MSE/Soil Nail walls using numerical analysis and limit equilibrium approaches. *HBRC Journal, 12*(1), 63–70. https://doi.org/10.1016/j.hbrcj.2014.06.012.
17. Galve, J. P., Cevasco, A., Brandolini, P., et al. (2016). Cost-based analysis of mitigation measures for shallow-landslide risk reduction strategies. *Engineering Geology, 213*(Supplement C), 142–157.

Pesanggrahan River Management in the Recent Times, the Anthropocene Era: A Case Study of Sangga Buana Urban Forest, Jakarta

Neli Triana, Linda Darmajanti, Yophie Septiady, Khoirunurrofik, and Komara Djaja

Abstract In recent times, actually humans live in the Anthropocene period. The river and its flow area, part of the city space, are also affected. This study is to collect a set of strategies for the integration of the active role of local communities and government policies in managing the river and making the river remain sustainable for human needs now and later. The study was conducted in Sangga Buana Urban Forest, Karang Tengah, Lebak Bulus, and South Jakarta. The management of the banks was carried out by the local community, the Kelompok Tani Lingkungan Hidup/KTLH Sangga Buana. One of the research problems was tracing the process of the emergence and development of urban forest, the pattern of relations between the KTLH and the government in explaining the existence of the urban forest management community. The other was uncovering the pattern of network and actor formation, division of labor and value system, as well as responsibility in the pattern of these relations. This research was a qualitative research, in which primary data were obtained through field observations and in-depth interviews and secondary data were from related publicly information. In conclusion, human and non-human factors have a large share in the formation of urban forest. The sustainability and management of urban forest show the pattern of relations formed between the KTLH, the government, and other parties. However, the relationship was only limited to mutual use and the Anthropocene concept has not been realized yet.

Keywords Anthropocene · Sustainable · Urban forest · River management

N. Triana · L. Darmajanti · Y. Septiady · Khoirunurrofik (✉) · K. Djaja
Urban Development Studies, School of Strategic and Global Studies, Universitas Indonesia, Depok, Indonesia
e-mail: khoirunurrofik@ui.ac.id

Khoirunurrofik · K. Djaja
Department of Economics, Faculty of Economics and Business, Universitas Indonesia, Depok, Indonesia

S. S. Emamian et al. (eds.), *ICSDEMS 2019*,
https://doi.org/10.1007/978-981-15-3765-3_15

1 Introduction

The river is one part of the natural environment that is very affected by human behavior. This includes rivers in Indonesia as well as Jakarta. According to Gunawan [1], Jakarta is still accustomed to river management approach limited to canalization to overcome flooding. Managing canals does cure Jakarta's floods for now. However, if there is new energy, i.e., higher rainfall with a longer period, the danger and loss will be greater because the capacity of the water is exceeded. According to Gunawan, what happened in the Dutch era was now repeated by the central government and the Provincial Government of DKI Jakarta. The river also becomes very small for humans, for the city, and for life, when it is only associated with flooding. This reduces the role of the river as a source of life and civilization. For that reason, in the contemporary context, a different approach is needed in managing rivers, including rivers in Jakarta. A unification of understanding about the current condition of the rivers and how large humans influence the river is needed. In Rivers of the Anthropocene book [2], it is said that humans will forever depend on the river because the river is a source of water and many other important activities depend on it including maintaining the balance of the earth.

Understanding Kelly's description, an interesting statement was obtained, that is, when talking to a river, it is not only seeing the flow. River entities include river banks, which follow along the flow from upstream to downstream, and areas affected by the river flow which are referred to watershed (Daerah Aliran Sungai/DAS) as a whole. In the context of Jakarta, the city that has 13 rivers and its tributaries, the management of these rivers means the management of the city itself. Here, the case of the Sangga Buana Urban Forest became interesting to study because from the information and preliminary data collected, this area of banks, which was very lush green, was managed by the local community since the 1990s. The Kelompok Tani Lingkungan Hidup/KTLH acknowledged that the land they managed was not their property, but the property of the state. Concerning the problem of my research, the management system of the Pesanggrahan River, especially in the Sangga Buana Urban Forest, would be uncovered by using concepts that refer to the Anthropocene concept to analyze the field findings.

The above concepts built and strengthened the Anthropocene concept: how the government and society should act in the current era, the Anthropocene era. Three things that would be answered in this study are based on the analysis of field findings using the concepts above: firstly, the emergence and development processes of Sangga Buana Urban Forest, secondly, the pattern of relations between KTLH, the government, and other parties in the management of Sangga Buana Urban Forest, and thirdly, the networks and actors in the establishment and management of Sangga Buana Urban Forest.

This research was conducted to be able to answer the research problem above. It was hoped that recommendations could be obtained for the proper river management pattern, especially the management of the Pesanggrahan River in the Anthropocene era, and become a gap for further research. The focus of this study was to examine the mindset and behavior of the actors who underlain river management. The mindset and behavior of these actors were studied holistically in a series of networks of the community, which illustrated the basis of the actors' thinking and actions. The series of community networks reinforced the limits of power and control of space in river management, the management of Sangga Buana Urban Forest in the Pesanggrahan River in the context of river management in the Antropocene era. Therefore, specifically the focus of this study was on the community study, which was limited to the problem of structures and community networks in the context of river management in the Anthropocene era.

2 Literature Review

2.1 River, Community, and Anthropocene

The river and its watershed, which have been the center of occupancy and the emergence of human civilization, have become two fields which are studied seriously in the concept of Anthropocene. In Rivers of the Anthropocene, Kelly [2] describe the case of Tyne River, Newcastle, UK. Syvitski [3] said that the concept of Anthropocene emerged and developed long ago. Until now, the scientists in various fields still debate and try to perfect it. However, at least from 1990, the Anthropocene concept among scientists began to narrow. There is a kind of claim not to stop interpreting change on this earth only as a result of global warming. Furthermore, Crutzen and Stoermer [4] state that it is necessary to develop a strategy to realize sustainable ecosystems, which can be widely accepted by the public, a strategy which can counteract human influences that have a very massive impact on the earth. Crutzen asserted that more research was needed and wisely applied the results of the research and invited the society as part of the research and embraced the social communities. This is to jointly seek the best path, which is still able to fulfill human needs and sustainable ecosystems.

Suparlan [5] explained the fundamental meaning of the community. According to his study, the community is understood as a smaller unit of life than the society, occupying and living in a certain area with unclear boundaries. This distinguishes the community from the society with the clear boundaries. Even so, the members of the community know each other through various social networks. They can be bonded together through networks of kinship and marriage. The definition of the community from Suparlan is still relevant to be used in the current era when information flows and technological advancements enable interpersonal individuals to interact even though they do not have the same attachment to the same area of life and kinship. Referring to Suparlan's statement, Septiady [6] stated that community conceptions make more city citizens more concerned about development facilities and the environment around their lives. To achieve this, strategies are needed in every design of urban spaces, that is, when the community is positioned as the subject of development. Thus, the patterns of urban space are formed in accordance with the expectations of interaction and communication from its citizens, not the hope of a handful of parties benefiting. This research looks at how the community was not limited to internal Sangga Buana KTLH as the manager of the Sangga Buana City Forest. The social relations among the KTLH, the local government, and other parties in the Sangga Buana Urban Forest area were also expected to be revealed.

The originator of the actor–network theory (ANT) stated that the Anthropocene concept whose idea was first thrown by geologists would inevitably affect all types of science/discipline. They have to think again. One could not be just looking at their side; thus, transdisciplinarity is needed. He stated, how can "normalize" anything that exists but at the same time the "natural strength" has been transformed by involving the role of humans in each piece. Humans are called a single role as well as contributing in many roles and connecting to each other in changing the earth so that the past form is difficult to recognize. The actor–networking theory basically attempts to explain that there is a balanced role between human subjects or humans with non-human subjects, namely nature or the natural environment. However, it is known from several cases involving the relationship between the community and the natural environment that the leadership greatly determines the success or failure of the community to transform.

More significant issues related to river management are offered by Chou [7]. Using the ANT framework, the results of Chou's research found that river management was identified as a relation product of key actors in the network who led the movement of a series of sources as tools, technology, materials, money, and triggered related actors to play roles that had been designed by the dominant actor. Researchers focused on the process of negotiation, representation, and displacement that affected the development of relations between actors and entities. In building relations with entities, the actors gave identity to each entity, as well as interests, roles, and actions that must be followed, including projects that must be done. Furthermore, there was a transformation until a network was formed between communities and the government and other parties. This is shown in Chou's research on the management of the Keelung watershed, Taipei, Taiwan. By using the basis of the actor–network theory (ANT) framework approach like Chou, the actors who played a role in the Sangga Buana Urban Forest were explored. The roles of non-human actors, namely rivers and green areas that are managed into urban forest, as well as human actors from the community, provincial-level local governments to villages, as well as other parties involved in urban forest management were revealed.

Furthermore, researchers focused on the process of negotiation, representation, and displacement that affected the development of relations between actors and entities. In building relations with entities, actors gave identities to each entity, also interests, roles, and actions that had to be followed, as well as projects that had to be done. Then, there was a transformation until a working network was formed. The transformation went through four stages:

a. Problematization—translating what can potentially be a common goal and how to present the problem faced as well as the desired solution and becoming the goal and solution of all entities
b. Interessement—all entities were involved in the project and blocked other flows/opposing groups
c. Enrollment—the role of actors was defined and distributed according to a certain level
d. Mobilization—the stage when the representation of the previous three stages was realized and transformed. Even so, the transformation that often occurred overlapped between stages

2.2 Basic Rules of Urban Green Open Space

Referring back to the case examined in this study, it was necessary to understand the concept of urban forest in accordance with the rules that apply in Indonesia and in particular Jakarta, where the Pesanggrahan River and Sangga Buana Urban Forest exist. There are three main rules regarding spatial planning which regulate the provision of green and non-green open spaces in urban areas. Urban forest is included in one type of green open space. The three main rules are Law Number 26 of 2007 concerning Spatial Planning, Government Regulation Number 26 of 2008 concerning National Spatial Planning, and Regulation of the Minister of Public Works No. 05/Prt/M/2008 concerning Guidelines for Provision and Use of Green Open Space in the Urban Area. Referring to these rules, the green open space (*ruang terbuka hijau*/RTH) is needed in a city which aims to maintain the availability of land as a water catchment area; create urban planological aspects through a balance between the natural environment and built environment that is useful for the benefit of society; and improve the harmony of the urban environment as a means of protecting the urban environment that is safe, comfortable, fresh, beautiful, and clean.

Urban forests, in this case, can be said occupied the highest position in the level of green open space: the highest level in terms of its function in the nature conservation. For this reason, making an area as an urban forest must meet many requirements, ranging from green cover and the types of trees or plants that must exist, minimum area, and others. For the interactive park RTH, for example, it does not have to be as tight as the urban forest in terms of its green cover. Interactive parks tend to be open spaces that provide more portions for playing facilities or joint activities between residents to interact with each other.

3 Research Method

The map of the research location can be seen in Figs. 1 and 2.

Fig. 1 Location map of Sangga Buana urban forest [8]

Fig. 2 Location of Sangga Buana urban forest. *Source* Google Map and Earth

This research applies mixed methods utilizing primary and secondary data. Primary data were obtained directly from the field in Sangga Buana Urban Forest, from informants. The informants were selected based on certain criteria: the main figure of KTLH Sangga Buana, core members, volunteers from residents and non-residents, as well as representatives from the government in the village, sub-district, agency, and central government levels which were directly related to the river management, especially Pesanggrahan River. Primary data were obtained from in-depth interviews with community leaders, senior community members, non-permanent community members, the government represented by the head of the department in charge of urban forests (Dinas Kehutanan DKI Jakarta), representatives of the sub-districts where the communities were located (Kecamatan Lebak Bulus dan Kelurahan Lebak Bulus), and the Head of the River Basin Center Ciliwung Cisadane (BBWSCC) who has the authority to manage rivers in Jakarta, including the Pesanggrahan River. In-depth interviews were conducted in stages during May 9, 2018–May 26, 2018.

Secondary data were obtained from the document Law Number 26 of 2007 concerning Spatial Planning, Government Regulation Number 26 of 2008 concerning National Spatial Planning, and Regulation of the Minister of Public Works No. 05/Prt/M/2008 concerning Guidelines for Provision and Use of Green Open Space in the Urban Area. In addition, there were some maps and urban forests/green open spaces/parks data from Dinas Kehutanan DKI Jakarta, Kecamatan Lebak Bulus, Kelurahan Lebak Bulus, and BBWSCC. Secondary data were also data from related offices or agencies, books, journals, theses, dissertations, and other literature sources. The primary and secondary data collected then were presented in the findings in the field. Additional information was also obtained from the news in the mass media related to the urban forest of Sangga Buana and its communities from 2013 until the study was conducted in May 2018.

After exposure of the findings, it was time for the analysis phase by using concepts as described previously. With a tool analysis in the form of theories related to river, community, Anthropocene, and actor–network theory to analyze primary data from in-depth interviews, it shows that there are some conflict potentials and gray areas in the relationship between KTLH Sangga Buana and the government. Reinforced with secondary data, the gray area seems increasingly clear and becomes the basis for drawing conclusions.

4 Results and Discussion

4.1 *Non-human Subject*

4.1.1 Pesanggrahan River

In relation to the Anthropocene era, it was evident that human intervention on the Pesanggrahan River had occurred since about the fourth century until now. The natural downstream of the Pesanggrahan River did not even exist because it was disodetected and replaced with man-made Cengkareng Drain. Various projects in the past until normalization in the present proved that Pesanggrahan River had been greatly influenced by humans. At present, the condition of the river was still bad and became one of the sources of floods in Jakarta and the surrounding areas. In most areas of the Pesanggrahan River that crossed Jakarta, rivers were also still treated as landfills and various other wastes. Part of the normalization project, which paved the sides of the river to become a kind of high fence around 2–3 m, further separated the river with humans inhabiting its banks. The river was increasingly positioned as an unsafe and wasteful place. This negated the position of the river where it should be. As stated by Chou [7], no matter how the current conditions, the river still exists. Increasingly urged and treated badly, the threat of potential disasters was getting closer. The condition of the Pesanggrahan River also encouraged changes in human behavior toward it. If not, humans themselves would bear the impact of the bad condition of the river.

4.1.2 Sangga Buana Urban Forest

In the midst of the bad conditions of Pesanggrahan River and the flowing area, there was a green area that existed. The existence of this green area contrasted with the condition of the river and was a driving factor that triggered changes in the behavior of local residents toward the river. The existence of this green area then triggered local residents to take the initiative to manage it and named it Sangga Buana Urban Forest. The condition of the city forest was well preserved. From the field observations, it was known that the trees there were quite lush; the banks of Pesanggrahan River in this area were beautiful; there were enough green open spaces and the residents could access them freely any time. This urban forest was the entrance for residents who had felt the benefits of the riverbanks which were minus the occupational building to keep it. In addition, the existence of the city forest was pushing the expansion of similar green areas. Moreover, the local residents felt that they could do any activity, from mere entertainment to what brought economic value in the green area.

However, it turned out that the status of the new urban forest was just unilateral pressure from the residents or in this case KTLH Sangga Buana. The green area was officially recognized by the government as an interactive park. The status of the green area in Karang Tengah had become unclear. On the one hand, there was a desire to make it a conservation area with the realization of the status of urban forest, but various parties, both the government and local residents, continued to enjoy the status of interactive park for activities that benefited them. This gray status was also indicated by the difference in the land area of the interactive park recognized by the Provincial Government of DKI Jakarta which was 2.7 ha. KTLH Sangga Buana, that stated the area as urban forest confirmed that the area of green land had reached 49 ha with 5 ha of which belonged to the Provincial Government of DKI Jakarta.

4.2 Human Subject

4.2.1 Community

Looking back at KTLH Sangga Buana, it should be emphasized that the KTLH was originally a group of citizens who later developed to remain a group and community. This refers to Septiady's opinion which refers to Suparlan that the conception of the community is more to bring the citizens to be more concerned about development facilities and the environment around their lives. There are strategies that are applied by citizens as the determinants of urban space design. The pattern of urban space is formed according to the interaction and communication of its citizens.

The strongness of Bi's personal network, the solid community of KTLH Sangga Buana under Bi, as well as supporting factors such as the proximity of location and kinship and the existence of Pesanggrahan River and green areas in Karang Tengah triggered the emergence and formation of Sangga Buana Urban Forest. Next, there was not a close relationship but mutual benefit, allowing urban forests and all activities of KTLH Sangga Buana to continue.

The existence of non-human subjects was supported by human subjects, especially a group of Karang Tengah residents who had the motivation to have a better environment to live, preserved the river and got economic benefits. This group of residents with a strong leader had a strong sense of belonging to the area where they lived. Motivation as the goal of the activities and the existence of non-human subjects, which were also strong, became a very powerful driver to realize the river and green areas which were then called urban forest by local residents. The sense of belonging and activeness in managing urban forests according to their desires and abilities created a sense of right and obligation to manage the area. Even though they realized that it was not their official property, because the land ownership of the riverbank was believed to be the right of the state or government, central or local, they were very confident of taking the role of controlling the city forest area. On the other hand, Bi as the grandson of the local community leader with the close relationship with the local community, both the members of the KTLH and volunteers, supported Bi as the leader of the community movement. Moreover, they were both from the Betawi ethnic background; some of them still had kinship, and they were residents who had been handed down for generations in the Karang Tengah area. These conditions strengthened their position to take a role in the control and power in Karang Tengah and its surroundings.

Various motivations which were the goal of this residents thickened and could integrate into a joint movement thanks to a central figure, Bi, as the initiator of the conservation of the Pesanggrahan River in Karang Tengah and the main driver for the formation of the Sangga Buana Urban Forest. Bi reflected the dominance of personal network which was the main determinant of the realization of the Sangga Buana Urban Forest. The explanation of the existence of non-human and human subjects (especially related to KTLH and Bi) also answered the formation process of Sangga Buana Urban Forest, according to the first research problem. In addition, there were supporting factors and other barriers that affected the formation and management of urban forest, as explained as follows:

4.3 Power and Control

From the concept of power and control as described by Septiady (2007 in Setiady 2017), BI's personal networks could not only carry out the "control" function of space. The government, both at village and center levels, as the holders of "power" was weakened by the strong "control" of Bi through factual performance and power symbols that he had with KTLH Sangga Buana as well as networks with the private sector. As a result, "power" was like being in Bi's hands and supported by KTLH Sangga Buana because this group could act to determine their own steps without the need for approval or permission according to the law from the government.

Although there was a prominent difference between Bi and his group and the government in seeing and enforcing Sangga Buana Urban Forest, this difference was as if it had never tried to be resolved or seen as a problem that needed an immediate solution at this time. Government relations, especially the Provincial Government of DKI Jakarta with KTLH Sangga Buana were tenuous. The local governments from the provincial, sub-district, and village levels were not trying to enforce the rules. They also did not attempt to at least rectify the naming or equalization of the green area according to the legal basis recognized by the government. The government did not want to declare KTLH as illegal. In addition, the government even admitted the benefits of the existence of Sangga Buana Urban Forest as one of the big contributing factors in adding the value of the Adipura Cup. Moreover, because of the Sangga Buana Urban Forest, Adipura became a gift subscription received by South Jakarta.

Bi and his group, as well as the government, were seen to let the current conditions continue. Both parties shared the benefits from the circumstance without legal status for the urban forest. Bi and his group, as well as the government, according to game theory [6], both were playing at the design level. At the design level, both parties were aware of each other's conditions, and what was best for both parties. They chose not to attack each other but also not to cooperate closely. It is called not to establish close cooperation because basically there is still a reciprocal relationship but there is no official cooperation between the two parties. Local government leaders also have good relations with Bi, sometimes visiting urban forests, as well as national leaders. There was an impression, reporting in the mass media about Bi and Sangga Buana Urban Forests that made the public and the figures encourage the realization of the role of power and power of Bi and KTLH. However, Bi and KTLH also used the law regarding the arrangement of the area and river management which was the product of the government and believed that they were citizens who had to defend or support their country, which was their foundation to manage the river banks. They, especially Bi, continued to use the phrase that river banks according to the rules should not be occupied and became land for massive buildings. However, the use of river banks, which was limited to everyday activities such as farming and other things that were not permanent and did not erode tree stands, was allowed to do.

4.4 Actor Network

In order to explain the formation process of networks and actors in the formation and management of Sangga Buana Urban Forest, in this section, an analysis, which discusses the stages of the process leading to transformation or pre-transformation, would be carried out. Pre-transformation stages were negotiation, representation, and displacement. Furthermore, the transformation process included four stages: problematization, intervention, enrollment, and mobilization. The results of the analysis in this section would converge on whether the formation and management of Sangga Buana Urban Forest were ideal according to the needs of river management in the Anthropocene era or vice versa.

As Chou's study of the Keelung River in Taiwan, as the dominant actor, according to the results of the analysis in the previous section, Bi was a key actor who was able to give roles to other actants. The other actants, which were the official members of KTLH Sangga Buana and volunteers, were also able to establish mutually beneficial relationships with the government even though it did not cooperate intensely. Bi was a key actor in the process of pre-transformation stages, namely negotiation, representation, and displacement. Looking at the beginning of the formation of the Sangga Buana Urban Forest, it could be seen that the main driver was from the internal side of the group and the community of KTLH Sangga Buana with Bi as their leader. For this reason, the pre-transformation stages were believed to be more involved in the internal KTLH Sangga Buana as a group or community.

Bi was able to concoct efforts to conserve the river with activities that were of economic value to the residents and make Karang Tengah residents become masters in their own environment. Here, Bi showed that non-human and human subjects were equally important, and could go hand in hand and mutually beneficial. This step complemented the representation and displacement processes that strengthened the internal position of the group and the community of KTLH Sangga Buana which became the provision for transforming the management of the Pesanggrahan River and the green area of its banks in Karang Tengah. This was because there were bigger challenges that would be faced in the transformation process. For example, when KTLH Sangga Buana had to deal with the power of the government and had to influence other parties, such as private developers.

In the transformation process, in the problematization stage, what Bi and KTLH Sangga Buana did was to reaffirm the common goal and present the problems and solutions in the form of urban forest management and economic activities. The aim of reaffirming their goal and efforts was to gain support from other parties, especially the government, the private sector, and the public in general. Bi and KTLH Sangga Buana needed the other parties to know their existence and purpose to realize that their urban forest dream basically needed the support of other parties and legal basis.

The next transformation process was enrollment and mobilization between Bi-KTLH Sangga Buana and the private sector to continue smoothly. Public support was shown by the positive news that had continued to exist over the past years from various mass media on various platforms. Private support was also strong, as shown by housing developers that had established intense cooperation at least in waste management and cooperation with other private parties in the form of cooperation in training activities and procurement of facilities in the Sangga Buana Urban Forest. However, one of the efforts to establish cooperation with the government could be called not fully successful whereas the government also benefited from the existence of Sangga Buana Urban Forest, such as at least not having to carry out normalization projects with dredging, widening, and concrete plastering in the Karang Tengah area which certainly reduced the burden on the central and regional budget.

5 Conclusions

In general, it can be concluded that river management that prioritizes the preservation of rivers while still accommodating the needs of today's urban communities, according to the principle of river management in the Anthropocene era, is very possible. Therefore, community activity is needed, especially those who live on the edge or close to the river and are directly affected by whatever happens in the river, to care and play a role in managing the river, while the government firmly takes the attitude and actions as law enforcers and holders of river management policies. However, there needs to be clarity and transparency in the vision, mission, organization, and budgeting in the collaboration between the community and the government in managing the river according to the demands of river management in the Anthropocene era.

Thus, what can be concluded in the case of Sangga Buana City Forest management are: (1) KTLH Sangga Buana is a community, (2) they had been the subject of city space development, in this case in the green area in Karang Tengah and managed it into Sangga Buana Urban Forest, (3) the data obtained show the number of activities carried out by KTLH Sangga Buana and many aspects surrounding it; this shows that the KTLH had already rooted strong and big, (4) there were lots of activities involving a lot of people, starting from the core members of KTLH Sangga Buana to volunteers from residents and non-residents; not only to demonstrate that the KTLH had strong root and big influence in the Karang Tengah area, especially in managing the river area and its banks, but it could also be part of the residents' resilience technique against area whose legal status was still gray, and (5) the relationship of mutual benefit was not the result of transformation but their respective ways to endure their conditions. In addition, this actually obscured Bi's position as the dominant actor because the holder of authority and law remained in the hand of the government, and river management, according to the demands of the Anthropocene era, had not been fully realized.

This study offers a recommendation for KTLH Sangga Buana that there should be clearer and more transparent management while remaining community-based regeneration of organization leader and organizing. In addition, transparency needs to be adopted in the financial accounting of the community. If needed, periodic audits by independent team from outside the community can be conducted. While, for the government, it is necessary to clarify the role of the government in the central level (Balai Besar Wilayah Sungai Ciliwung-Cisadane/BBWSCC), province, city, sub-district, and village as well as RT and RW in managing the rivers in the Anthropocene era. It is necessary to equate the vision and mission of river management in the Anthropocene era nationally to the regional and village levels; understand the rights and obligations of the government apparatus in carrying out the mandate of the law. Furthermore, it clarifies the role of the government in the collaboration between the government and the community in river management in each area where the community operates.

Furthermore, it is suggested to increase research related to environmental management in the Anthropocene era, including the management of cities and rivers. Therefore, studies related to diverse cases will be the important data contributors to help find patterns of the environmental management while still ensuring its sustainability and the fulfillment of the needs of people living in the environment. Finally, the cooperation between academics, government, and society both in the stages of researching and evaluating as well as determining the right river management in the Anthropocene era, especially in urban areas in Indonesia, is needed. At present, similar research is still very limited. A transdisciplinary collaboration is indeed a demand to be realized in this Anthropocene era.

References

1. Gunawan, R. (2010). *Gagalnya Sistem Kanal*. Jakarta: Penerbit Buku Kompas.
2. Kelly, J. M. (Ed.). (2017). *Rivers of the anthropocene*. University of California Press.
3. Syvitski, J. P. M. (2012). *Anthropocene: an epoch of our making*. Global Change Newsletter, The International Geosphere–Biosphere Programme (IGBP). A study of global change of the International Council for Science (ICSU). First Published in IGPB's Global Change magazine. No/Issue 78, Maret.
4. Crutzen, P., Stoermer, E. (2000). *The anthropocene*. Global Change Newsletter, The International Geosphere–Biosphere Programme (IGBP). A study of global change of the International Council for Science (ICSU). No. 41. May, pp. 17–18.
5. Suparlan, P. (2004). Masyarakat dan Kebudayaan Perkotaan: Perspektif Antropologi Perkotaan. Penerbit: Yayasan Pengembangan Kajian Ilmu Kepolisian (YPKIK).
6. Septiady, Y. (2011). Teori "Game" Dalam Memandang Green City Sebagai Usaha Penataan Kawasan Kota. Seminar Nasional Green City, Fakultas Teknik Arsitektur Universitas Pancasila, Jakarta, 5 June 2007 dan Seminar Nasional Green Psychology in Urban Setting, Fakultas Psikologi, Universitas Pancasila, September 14, 2011.
7. Chou, T.-L. (2012). ANT analysis on the river management of urban Taiwan: a case study of Keelung river in Taipei basin. Scientific Research JANT analysis on the river management of urban Taiwan: a case study of Keelung River in Taipei basin
8. Kompas H (2016) Penataan Sungai Kian Mendesak. Pembenahan Masih Berjalan Sepotong-sepotong. Monday, 29 Agustus 2016, pp 1, 15.

Simulation of Damage Due to Alkali–Silica Reaction in a Concrete Model at the Macroscale Level

Zarina Itam, Agusril Syamsir, Salmia Beddu, and Nur Liyana Mohd. Kamal

Abstract The rapid deformation caused by alkali–silica reaction may cause major problems in concrete structures. Main factors that affect ASR include the relative humidity, alkali and silica content, temperature, and porosity of the concrete. The vast numbers of influencing factors make ASR relationship a complex phenomenon to understand. Investigation of the mechanical deformation of the structure can be achieved by applying the theory of continuum damage mechanics. Prediction of the physical and chemical behavior of a structure can be achieved using damage mechanics, making it an appropriate method to study the behavior of the structure under the influence of alkali–silica reactivity. Hence, solution of the damage model as well as simulation of the ASR phenomenon is critically needed. In this research, an engineering example of a thermo-chemo-hygro-mechanical model of a concrete gravity dam at the macroscale will be studied for varying environmental conditions of temperature and relative humidity using the finite element method. Investigations found that temperature as well as relative humidity influences the latency and characteristic time constants, which indicates that the ASR expansion rapidity is dependent on the heat and moisture diffusion lengths into the structure, causing heterogeneous damage along the cross section of the macroscale structure according to the temperature distribution as well as relative humidity.

Keywords Alkali-silica reaction · Concrete · Gravity dam · Finite element

1 Introduction

Damage to concrete due to alkali–silica reaction or ASR is a phenomenon that was first recognized in the USA in the 1940s by Stanton (1940) and has since been observed in many other countries. Since then, many studies on that matter have been published. Factors that affect ASR vary greatly although it is unanimous that ASR occurs between deleterious silica from aggregates and hydroxide ions in the

Z. Itam (✉) · A. Syamsir · S. Beddu · N. L. Mohd. Kamal
Department of Civil Engineering, College of Engineering, Universiti Tenaga Nasional, Jalan IKRAM-UNITEN, 40300 Kajang, Selangor, Malaysia
e-mail: iZarina@uniten.edu.my

S. S. Emamian et al. (eds.), *ICSDEMS 2019*,
https://doi.org/10.1007/978-981-15-3765-3_16

pore solution that result from cement hydration. Other factors include the relative humidity, temperature, and porosity of the cementitious matrix.

What sets apart ASR from other concrete damage models is its heterogeneity, occurring at different concrete regions at different rates depending on the concrete composition as well as external influences, making predicting its behavior somewhat difficult. But despite the fact that ASR initializes in the mesoscopic regions of the concrete, the accumulative effects of its expansion escalate onto the macroscale level with the development of web cracking on the concrete surface [1]. Macroscopic material models are characterized by considering the heterogeneities and structural defects in an averaged sense and are therefore regarded as a homogeneous material. In investigating the mechanical deformation of the structure, the theory of continuum damage mechanics proves to be a suitable method. Damage mechanics can be used to predict the physical and chemical behavior of a structure, making it an appropriate method to study the behavior of the structure under the influence of alkali–silica reactivity. Therefore, solution of the damage model as well as simulation of the ASR phenomenon at the macroscale level in order to provide better understanding, or even solving the problem, is critically needed.

Two main aspects to be considered in ASR modeling are the kinetics of chemical reactions and the mechanical damage caused by the resulting expansion. Li and Coussy [2] developed a thermo-hydro-chemo-mechanical model to study the behavior of ASR affected concrete based on the Biot–Coussy theory on mechanics of reactive porous medium. Capra and Bournazel [3] suggested the ASR gel content of silica and alkali.

Numerical simulation of stress anisotropy due to ASR is performed by taking into consideration thermo-activation of alkali–silica reactivity and its dependency on relative humidity. In investigating the mechanical deformation of the structure, the theory of continuum damage mechanics can be used to predict the physical and chemical behavior of a structure, making it appropriate to study the behavior of the structure under the influence of alkali–silica reactivity [4].

Assuming that damage at a given temperature and relative humidity has isotropic behavior, the determination of the local ASR reaction extent ξ is performed with a stable time integration method; for instance, the backward Euler method has taken into account the corresponding moisture and temperature profiles for each time step of the numerical procedure. The stress and damage states from the mechanical response are obtained from the rheological modeling coupled with the damage law and adding the effect of thermal strain, $\varepsilon^{\mathrm{th}}$, ASR strain, ε^{a}, pore pressure, P^{f}. The constitutive equation can be modified into (1):

$$\sigma = (1 - \omega)\left[C\left(\varepsilon - \varepsilon^{\mathrm{a}} - \varepsilon^{\mathrm{th}}\right) - P^{\mathrm{f}}\right] \tag{1}$$

where $\mathbf{C}$ is the fourth-order linear elastic material tensor, and ω denotes the scalar damage parameter [4]. Lund [5] stated that the total thermal strain increment could be summed up over the current and history temperature range:

$$\varepsilon_{ij}^{\mathrm{th}} = \left[\int_{T_0}^{T} \alpha_{\mathrm{T}}(T - T_0) \right] \delta_{ij} \tag{2}$$

The post-computation of the thermoelastic strain starts with the iteration of (2), resulting in thermal strain values, $\varepsilon^{\mathrm{th}}$. The numerical solution of the equation for temperature $T + dT$ can be solved using the backward Euler method:

$$\varepsilon_{ij}^{th}(T - T_0) = \varepsilon_{ij}^{th}(T) + \alpha_T(T - T_0)\mathrm{d}T\,\delta_{ij} \tag{3}$$

where α^{T} is a thermal expansion coefficient matrix, T is the temperature at a given point, and T_0 is the temperature at which the structure is free of thermally induced strain. ASR strain can be represented in an equation for free linear expansion ε^{a} as shown in (4):

$$\varepsilon^{\mathrm{a}}(t) = \varepsilon_{\infty}\xi(t) \tag{4}$$

where ε_{∞} is the asymptotic ASR expansion. Equation (4) shows that ASR strain evolution is only governed by the noninstantaneous kinetics of the chemical reaction. In a stress-free condition, the only unknown is the reaction extent, ξ. The reaction extent ξ is determined from:

$$\xi(t) = \frac{\varepsilon}{\varepsilon_{\infty}} = \frac{1 - \exp(-t/\tau_{\mathrm{C}})}{1 + \exp(-t/\tau_{\mathrm{C}} + \tau_{\mathrm{L}}/\tau_{\mathrm{C}})} \tag{5}$$

Ulm [6] defined τ_{L} as the latency time as the time needed for the initial expansion of the gel in the presence of water. τ_{C} is the characteristic time that is defined by the intersection of the tangent at τ_{L} with the asymptotic unit value of the reaction extent, ξ. τ_{C} and τ_{L} are thermally dependent and can be expressed in terms of absolute temperature (TK = 273 + T °C) and the corresponding activation energies that abide the Arrhenius concept: The variables U_{L} and U_{C} are the activation energies, which represent the minimum energy required to trigger the reaction for the time constants and were determined to be $U_{\mathrm{L}} = 9400 \pm 500$ K and $U_{\mathrm{C}} = 5400 \pm 500$ K [7]. The time constants τ_{C} and τ_{L}, tend to have different intensities for drying, sealing, humidity, and immersity in water conditions, proving that ASR reaction extent is influenced not only by temperature as can be seen in (5), but also by the relative humidity.

2 Modeling ASR at the Macroscale

Various experimental evidences for ASR deterioration can be found from literature, for example by Comi [8]. Comi developed a chemo-thermo-damage model of a gravity dam that evaluates the local evolution of ASR swelling governed only by temperature, resulting in a damage model that is fully decoupled from the heat-diffusion problem. Fairbairn [9] simulated the stress anisotropy of ASR swelling, taking into account thermo-activation and humidity dependencies.

The numerical example presented here illustrates the performance of a two-dimensional concrete gravity dam tested for thermo-hygro-chemo-mechanical loading using the finite element method. The model presented in this paper is performed at the macroscale, where the model consists of a single phase, which is concrete. Although the problems are not examined in full engineering detail, the porosity of the matrix explicitly is not simulated and the model is sufficient to demonstrate the workability of the algorithm applied in this research. This has been proven feasible by a series of validation tests conducted with benchmark examples in the following section.

3 Methodology

3.1 Modeling Parameters

A two-dimensional cross section of a concrete gravity dam measuring 14,009.5 mm width at the bottom surface and 4000 mm width at the top surface has been simulated using SLang programming software developed and used in Bauhaus-Universitaet Weimar [10, 11]. The total height of the dam is 20,000 mm. The finite element mesh for the cross section consists of 1681 nodes with 1600 elements as shown in Fig. 1. The elements used in the dam are four-noded plane elements with four integration points. The structure is restraint at the bottom surface in both $x1$ and $x2$ directions,

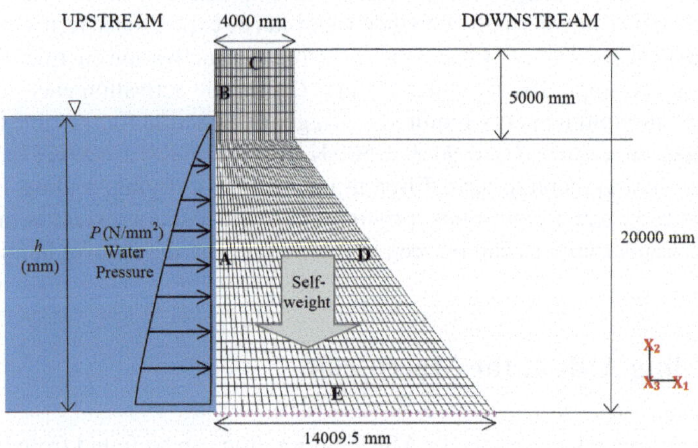

Fig. 1 Two-dimensional macroscale model of the undeformed concrete gravity dam and the acting boundary conditions

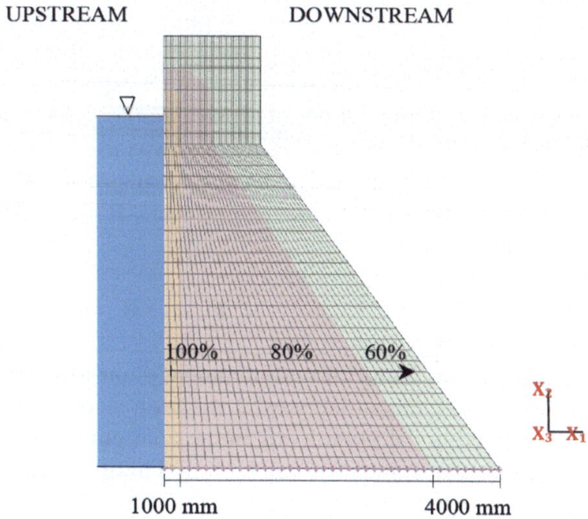

Fig. 2 Coloured zones depicting different relative humidity conditions ranging from 60 to 100%

while all other surfaces are allowed to move. The number of degrees of freedom for heat and pressure elements is 1681 with one degree of freedom per node, and the number of degrees of freedom for mechanically loaded elements is 3362 with two degrees of freedom per node to accommodate for the $x1$ and $x2$ directions, using plane strain in the simulation. Boundary conditions, as can be seen in Fig. 1, show that the upstream surface of the dam is exposed to water. This huge amount of body represents the mechanical loading for the system with the water pressure. Fluid flow in the dam will only be facilitated by fluid pressure from the upstream and gravitational force.

Relative humidity influences the latency and characteristic time constants, which will in turn influence the ASR reaction rate. In this research, the effect of relative humidity will be presented by different relative humidity values. Figure 2 shows the concrete gravity dam with three colored zones reflecting different values of relative humidity. Starting from the boundary, which is exposed to water, the orange zone has a relative humidity of 100%, which slowly reduces to 80% in the red zone, and finally in the green zone, the relative humidity ends at 60%.

3.2 Theories and Assumptions

Considering the fact that damage caused by ASR is due to expansion, behavior of the concrete gravity dam under compressive stress can be modeled as linear. Since the ultimate compressive strength value, f_c, is not exceeded, damage due to compression is assumed negligible. Creep and shrinkage will not be considered in this simulation. The material properties are taken to be constant and homogeneous throughout the macroscale domain. Assume that the latency and characteristics time values are taken as constant for an initial condition of 17 °C at all boundaries with a relative humidity of 80%. In modeling the effect of ASR in this research for a concrete gravity dam, the asymptotic ASR expansion, ε_{∞}, is taken as 0.22% [12]. As a means of studying the orientation of damage in the structure until fully damaged, the Newton–Raphson iterative method was used. Using this method, the difference between externally applied nodal point loads vector, $\mathbf{F}_{n+1}$, and the internal node point loads that corresponds to element stresses, $\mathbf{R}^i$, at iteration step i must equal to zero as shown in (6):

$$\mathbf{F}_{n+1} - \mathbf{R}^i = 0 \tag{6}$$

where $\mathbf{R}^i$ is the vector of internal nodal point loads in the configuration that corresponds to the element stresses at iteration step i, and $\mathbf{F}_{n+1}$ is the vectors of externally applied nodal point loads in the configuration at iteration step i. More on the Newton–Raphson iteration can be found in literature.

4 Results and Discussions

From Fig. 3, it could be seen that the displacements due to a varied relative humidity are larger at locations with higher relative humidity, which causes more rapid ASR reaction rate, for example, in the orange zone. This zone, having a relative humidity of 100%, expands faster when compared to the green zone, which has a relative humidity of 60%, which is the minimum optimal relative humidity condition for ASR reaction to initiate. This causes a high distribution of tensile stresses in that region, larger than the compressive stresses caused by hydrostatic pressure, and larger than the tensile stresses in the downstream region as reflected in Fig. 3. Seeing that the upstream region expands faster and with larger values than the downstream region, the dam deflects outwards and leans toward the downstream, causing compressive stresses in the vertical direction of the downstream region.

The damage contours for the varying relative humidity is also seen to be more localized to regions of higher relative humidity. Damage initiates at the zone where relative humidity is 100%, at the upstream region and continues inwards. There is no damage at the regions with the low relative humidity of 60%. However, under these conditions, ASR reaction takes a long time to occur and start causing damage to a structure. Damage for the constant conditions starts at approximately 3600 days, while damage for the varying relative humidity starts at approximately 1800 days, which is twice the time for the constant conditions case. However, after 10,950 days, damage in the constant conditions case is smeared throughout the dam, while damage in the varying relative humidity case is localized in the higher moisture regions.

Figure 4 shows the curve for ASR reaction extent against time for two relative humidity values of 80% and 100% for Nodes 126 and 513, respectively. The curve follows a sigmoidal shape, where at the start of the reaction, the ASR reaction rate, ξ, is 0, meaning that the reaction has not initiated. As the reaction continues, the

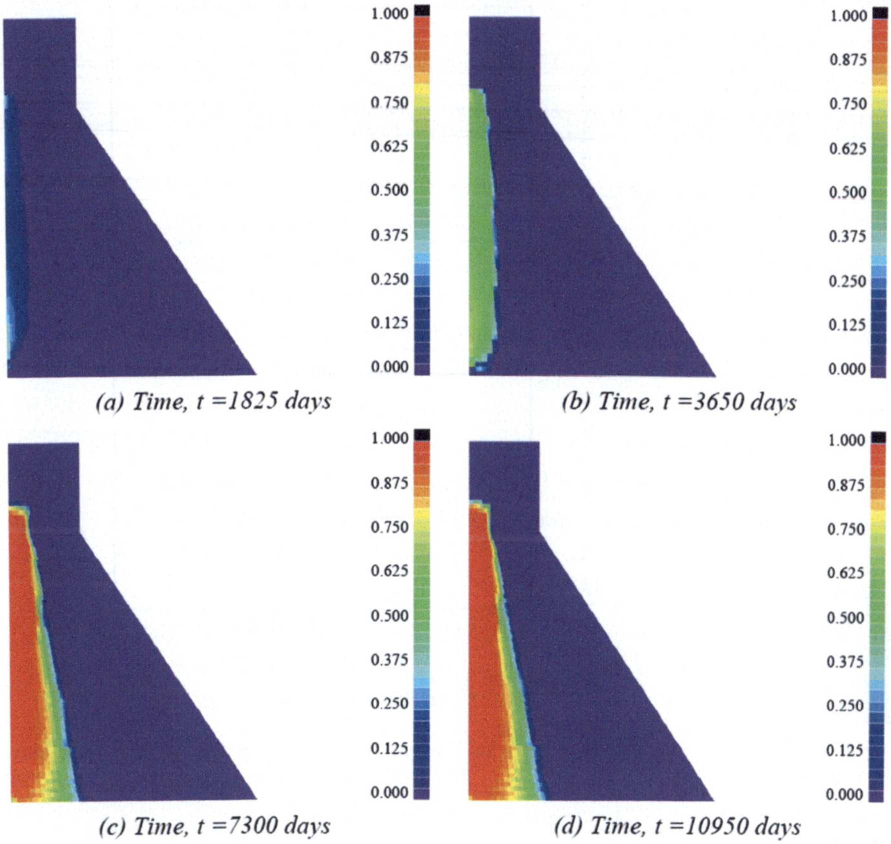

Fig. 3 Damage variable due to ASR and hydrostatic pressure for a constant temperature of 20 °C and relative humidity ranging from 60 to 100%

reaction rate increases, its rapidity depending on the temperature as well as the relative humidity. As can be seen, under such circumstances, the reaction rate for the relative humidity of 100% happens at a much faster rate than for the relative humidity of 80%. However, for both cases, ASR takes a long time until fully reacted, which enables us to conclude that even though ASR expansion may cause damage to a concrete structure, additional stresses to the structure accelerate the damage process. The exposure conditions of the structure also determine the orientation and initiation of the ASR damage.

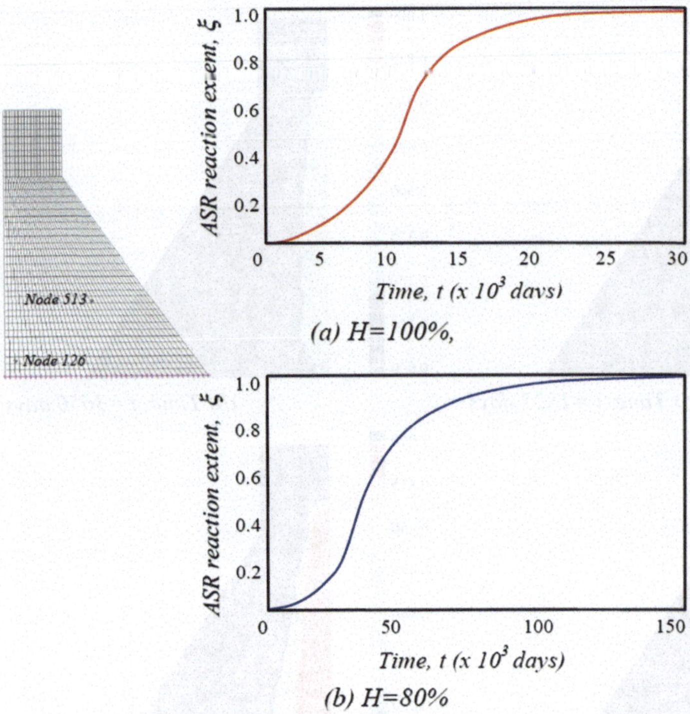

Fig. 4 ASR reaction extent against time for an isothermal condition of 20 °C and relative humidity of 80 and 100%

5 Conclusions

The aim of this research was to determine the effects of temperature and relative humidity on ASR expansivity in a concrete structure. This was performed at the macroscale level in order to determine ASR expansion mechanism on a smeared level, allowing us to determine its initiation and orientation.

The macroscale model was developed to have homogenous material properties for concrete. The determination of ASR expansion with inverse analysis from experimental results and application into modeling shows that ASR expansion can be replicated in numerical simulation, provided with the proper material properties and environmental parameters.

The engineering study of a macroscale concrete gravity dam incorporated with the effects of temperature, relative humidity, and mechanical loading cases as isolated and combined cases in this thesis shows that individually, temperature, relative humidity, and mechanical loading influence ASR expansivity and cause, its own affects to the expansion or contraction of the structure.

ASR has an increased effect due to moisture content. A higher relative humidity increases the characteristic and latency time constants, which means that a shorter time is needed before expansion occurs. As a benchmark unanimous among researchers, a minimum of 60% relative humidity was required for ASR deformation to occur.

Thermal difference leads to tensile or compressive stresses in concrete. Temperature, as well as relative humidity, influences the latency and characteristic time constants, which dictates the rapidity of ASR expansion, showing its dependency on the heat and moisture diffusion lengths into the structure, rendering heterogeneous values across the cross section of the structure according to the relative humidity and temperature distribution. Hence, a mesoscopic study would be more suitable to evaluate explicit ASR effects in terms of initiation and orientation in concrete structures.

Hydrostatic pressure causes compressive stresses in a structure. However, depending on the concrete material properties, boundary conditions, and environmental conditions, its effect could be rapidly overcome due to expansion of the structure from ASR and thermal expansion. The shape of the structure, as well as its restraints, also influences the deflection of the structure. The tensile stresses in the structure, due to thermal difference and ASR deformation, cause the dam to expand toward the region with lower thermal difference and ASR deformation.

Therefore, it can be concluded that the intensity of alkali–silica reactivity in a concrete structure due to ASR expansion depends on a lot of factors, most importantly the temperature and relative humidity. Other external factors have an influence on the expansion and orientation of alkali–silica reactivity which are the material properties, boundary conditions, and if applicable, external loading

Acknowledgements The authors gratefully extend the highest gratitude to Prof. Dr.-Ing.habil. Jochen Stark and the staff of Finger-Institut Baustoffkunde for the experimental data and Prof. Dr.-Ing.habil. Carsten Könke and the staff of Institut für Strukturmechanik of Bauhaus-Universität Weimar, Germany, for their guidance. The authors wish to acknowledge YCU Grant 201801001YCU/17 for the sponsorship for publication of this paper.

References

1. Itam, Z. (2015). Mesoscale simulation of alkali-silica reaction. *International Journal of Scientific & Engineering Research, 6*(5). May-2015 44 ISSN 2229-5518.
2. Li, K., & Coussy, O. (2004). *Comprehensive chemo-mechanical modelling of ASR expansion in concrete structures*. In *Proceedings of the 12th International Conference on Alkali-Aggregate Reaction in Concrete.*
3. Capra, B., & Bournazel, J.-P. (1998). Modelling of induced mechanical effects of alkali-aggregate reactions. *Cement and Concrete Research, 28*(2):251–260.
4. Itam, Z., & Husain, H. (2015). Modeling of alkali-silica reaction. In *A two-phased material model. UTM Jurnal Teknologi.* https://doi.org/10.11113/jt.v76.5637.
5. Lund, E. (2002). *Short note on thermal and thermo-elastic finite element analyses.* Aalborg University.
6. Ulm, F.J., Coussy, O., Kefei, L., & Larive, C. (2000). Thermo-chemo-mechanics of ASR expansion in concrete structures. *Journal of Engineering Mechanics, 126*(3).

7. Larive, C., Laplaud, A. & Coussy, O. (2000). The role of water in alkali-silica reaction. In *11th International Conference on Alkali-Aggregate Reaction*. Québec.
8. Comi, C., Fedele, R., & Perego, U. (2009). A chemo-thermo-damage model for the analysis of concrete dams affected by alkali-silica reaction. *Mechanics of Materials, 41*, 210–230.
9. Fairbairn, E. M. R., Riberio, F. L., Toledo-Filho, R. D., Lopes, L. E., Silvoso, M. M., Águas, M. F. F., & Guedes, Q. M. (2004). Smeared cracking FEM simulation of alkali silica expansion using a new macroscopic coupled model. In *Proceedings of the 12th International Conference on Alkali-Aggregate Reaction in Concrete*.
10. Itam, Z. (2014). *Numerical simulation of thermo-chemo-hygro-mechanical alkali-silica reaction model in concrete at the mesoscale and macroscale*. Bauhaus-Universität Weimar.
11. Stark J. (2008). *Alkali-Kieselsäure-Reaktion*. F. A. Finger-Institut für Baustoffkunde. Bauhaus-Universität Weimar.
12. Itam, Z., et al. (2018). Simulation of alkali-silica reaction model in a concrete gravity dam at the macroscale and mesoscale. *Science Direct Materials Today: Proceedings*.
13. Bangert, F., Kuhl, D., & Meschke, G. (2004). Chemo-hygro-mechanical modelling and numerical simulation of concrete deterioration caused by alkali-silica reaction. *International Journal for Numerical and Analytical Methods in Geomechanics, 28*, 68–714.
14. Bazant, Z. P.. & Steffens. A. (2000). Mathematical model for kinetics of alkali-silica reaction in concrete. *Cement and Concrete Research, 30*, 419–428.

Combination of LEM and FEM Analysis for Stability of Concrete Cantilever Retaining Wall

Rohaya Alias, Siti Jahara Matlan, and Nur Asmaliza Mohd. Noor

Abstract Concrete cantilever retaining wall is a structure used to retain the soil on the slopes and prevent the soil from sliding. It must be stable and strong to resist the forces acting on it. Normally, concrete cantilever retaining wall uses limit equilibrium method (LEM) in its design analysis. However, this method does not provide a comprehensive design analysis. It only involved an external stability check for sliding, overturning, and bearing capacity. As a result, the retaining wall is not enough stable and it often collapses. Hence, this study introduces the combination of LEM and FEM in the concrete cantilever retaining wall design analysis to produce the overall stability. The design analysis was carried out using PROKON software that involved LEM, and simulations were performed using SIGMA/W and SLOPE/W to generate FEM. The comprehensive design analysis was produced from the combination of both methods which consist of value of factor of safety (FOS) for sliding, FOS for overturning, FOS for bearing capacity, maximum foundation settlement, maximum surface settlement, maximum deflection, and global stability. This study can help the consultant engineers in producing the comprehensive and safe concrete cantilever retaining walls and reduce the problem of retaining wall failure.

Keywords Combination · Comprehensive · Design analysis · Retaining wall · Stability

1 Introduction

Retaining wall is a structure that is built to retain vertical or nearly vertical earth bank or any other material. It is also used to support foundation ditches and stabilize

---

R. Alias (✉) · N. A. Mohd. Noor
Faculty of Civil Engineering, Universiti Teknologi MARA Pahang Kampus Jengka, 26400 Bandar Tun Razak Jengka, Pahang, Malaysia
e-mail: rohaya_alias@uitm.edu.my

S. J. Matlan
Civil Engineering Programme, Faculty of Engineering, Universiti Malaysia Sabah, 88400 Kota Kinabalu, Sabah, Malaysia

S. S. Emamian et al. (eds.), *ICSDEMS 2019*,
https://doi.org/10.1007/978-981-15-3765-3_17

slopes. Retaining walls are widely used in various fields of engineering such as roads, harbors, dams, subways, railroads, tunnels, mines, and military fortifications [1].

Generally, retaining walls are broadly classified as gravity, semi-gravity, and cantilever retaining walls. Retaining walls may be constructed of masonry, concrete, or sheet piles [2]. The gravity walls are the oldest and simplest type of retaining walls. The gravity walls are heavy, thick, and stiff enough that they do not bend; their movement occurs essentially by rigid body translation or rotation. It utilizes entirely its own weight to produce the necessary stability. Semi-gravity retaining walls are intermediate between the gravity- and cantilever-type walls. Meanwhile, cantilever walls are bending as well as translating and rotating, which rely on their flexural strength to resist lateral earth pressure [3].

Among the concrete retaining walls, the cantilever wall is the most widely used as it is economical. The concrete cantilever retaining walls are used in basement of buildings, as abutments for bridges, as flood walls in irrigation works as well as for retaining ores, minerals, and other granular materials [4]. The concrete cantilever retaining walls are constructed usually shaped like an inverted *T* or *L* [5]. Generally, the height of the concrete cantilever retaining wall can reach up to 6 m. They consist of a relatively thin stem and a base slab. The base is also divided into two parts, the heel, and toe. The heel is the part of the base under the backfill. The toe is the other part of the base [6].

The determination of the stability of concrete cantilever retaining wall is an important task in geotechnical engineering practice. Concrete cantilever retaining walls utilize the weight of the soil itself to produce stability. Patil and Wagh [7] carried out a work on providing a pressure relief shelf toward the back side of the wall which increases the stability and overturning capacity of the wall. By providing shelf, quantity of construction materials such as concrete and steel is reduced by 35% and 18%, respectively. Padhye and Ullagaddi [8] have experimentally proved that providing the shelf to the back of retaining wall increases the stability against overturn and sliding.

The failure of concrete cantilever retaining wall is categorized as a geotechnical disaster because it causes major economic losses. According to Aswin [9], the main reason for the failure was the improper design of the concrete cantilever retaining wall. Retaining walls must be designed carefully to withstand lateral pressure of the earth, which tends to cause a failure of the structure. Design of concrete cantilever retaining walls involves stability checks for overturning, sliding, and bearing capacity [10].

Design of retaining walls has traditionally been carrying out using simplified method of analysis such as limit equilibrium method (LEM) and empirical approaches. This approach is limitative and lacks the information needed for an accurate design. Thus, they cannot provide the engineer with all the desired design information. Therefore, this paper presents a study to introduce the comprehensive and safe design analysis of concrete cantilever retaining walls using the combination of limit equilibrium method (LEM) and finite element method (FEM). The introduction of numerical or finite element software has resulted in considerable advances in the analysis and design of civil engineering structures. Therefore, it is of great encouragement to study and understand the use of this software in solving practical problems of concrete cantilever retaining walls.

2 Methods

In this study, design of concrete cantilever retaining walls was created by PROKON software using limit equilibrium methods (LEM). Figure 1 shows the concrete cantilever retaining wall layout. The design process was carried out in accordance with BS 8110 procedures and used Coulomb theory. Input data consists of (i) height of wall, (ii) angle of slope, and (iii) surcharge. Output data consists of the external stability, namely factor of safety (FOS) for: (i) sliding, (ii) overturning, and (iii) bearing capacity.

The finite element analysis was conducted using GeoStudio software, which is a widely used finite element program for a broad variety of geotechnical, geoenvironmental, civil, and mining engineering projects [11]. It is mostly used for analysis like stress strain, seepage, slope stability, dynamic analysis, and water drawdown in the reservoir [12]. It is also easy to learn and is user-friendly. In this study, the GeoStudio 2012 software consisting of SIGMA/W and SLOPE/W was used to generate FEM calculations. Using the same geometry of the wall and surcharges in LEM method, simulations were performed using a two-dimensional finite element analysis to calculate (i) maximum foundation settlement, (ii) maximum surface settlement, (iii) maximum deflection, and (iv) global stability. Table 1 shows information

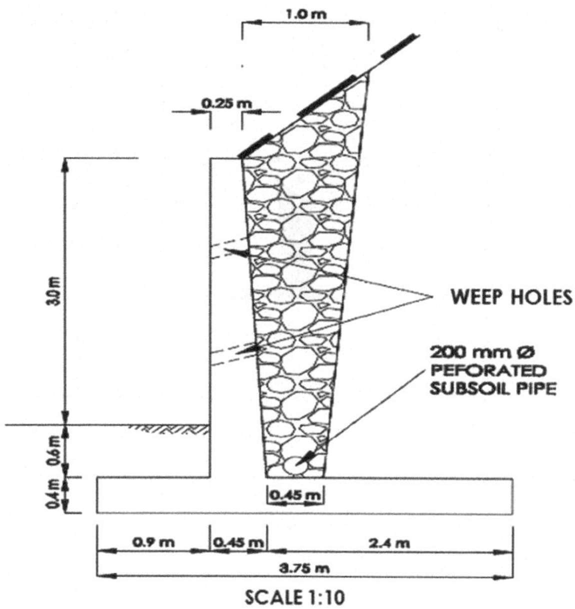

Fig. 1 Concrete cantilever retaining wall layout

Table 1 Properties of materials

Material	Model	Unit weight γ	Cohesion c	Frictional angle ϕ	Young's modulus E	Poisson's ratio v
		(kN/m^3)	(kN/m^2)	(Degree)	(kPa)	
Soil layer (clay)	Elastic-plastic (Mohr-Coulomb)	17	29.6	26.5	50×10^3	0.3
Concrete wall (concrete grade = M35)	Linear elastic	24	500	45	30×10^6	0.2
Backfill	Linear elastic	18	0	48	200×10^3	0.3

on the properties of materials used in FEM modeling. The unit weight values for concrete wall and backfill are similar to those used in LEM design. A mesh with an element size of 1.0 m and a total of 287 elements were applied in FEM analysis. Figure 2 shows the modeling of the concrete cantilever retaining wall. Then, the results of FEM analysis will be compared to the results of field study using the original design of concrete cantilever retaining wall structure. The comparison between the finite element analysis and field monitoring results would help engineers to get better understanding of the real soil behavior compared to finite element modeling.

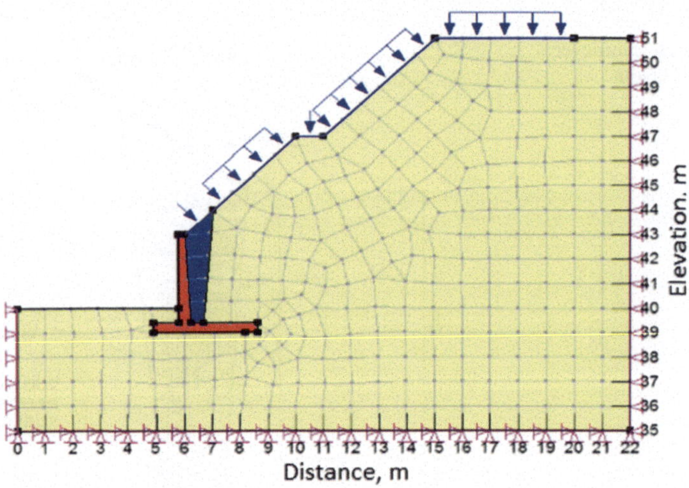

Fig. 2 Modeling of the retaining wall

Table 2 Results of LEM analysis

Stability checks	LEM
FOS for sliding	8.04
FOS for overturning	13.78
FOS for bearing capacity	4.07

Table 3 Results of FEM analysis

Stability checks	FEM
Maximum foundation settlement	11.77 mm
Maximum surface settlement	21.43 mm
Maximum deflection	4.34 mm
Global stability	2.342

3 Results and Discussion

3.1 *LEM Analysis*

The detailed result obtained from the LEM analysis is summarized in Table 2. The results of external stability, which consists of stability of sliding, overturning, and bearing capacity that developed from LEM analysis, are satisfactory.

3.2 *FEM Analysis*

Table 3 shows the summary results obtained from FEM analysis.

The maximum foundation settlement locations for FEM analysis are shown in Fig. 3a. It was found that the maximum foundation settlement was located at node 78 and 79 with a value of 11.77 mm. The maximum foundation settlement occurs at the back of the base of the wall. Figure 3b shows the maximum surface settlement using FEM analysis, with a value of 21.43 mm at node 124. It was found that the maximum surface settlement was located at the shoulder of berm drainage. Figure 3c shows the results obtained through the FEM analysis of the maximum deflection. The maximum deflection of the wall after eight months of construction occurs at node 157 with a value of 4.34 mm. Global stability that has taken into account the influence of settlement and deflection using FEM analysis is shown in Fig. 3d. The global stability was found with factor of safety (FOS) value of 2.342. This result indicates that the retaining wall has a satisfactory and non-critical global stability.

When the results of FEM analysis and field study are compared, it was indicated that the FEM simulation model is reliable because it produces results that are very similar to those obtained in the field. The comparisons elucidate that the developed finite element model is satisfactory in its performance.

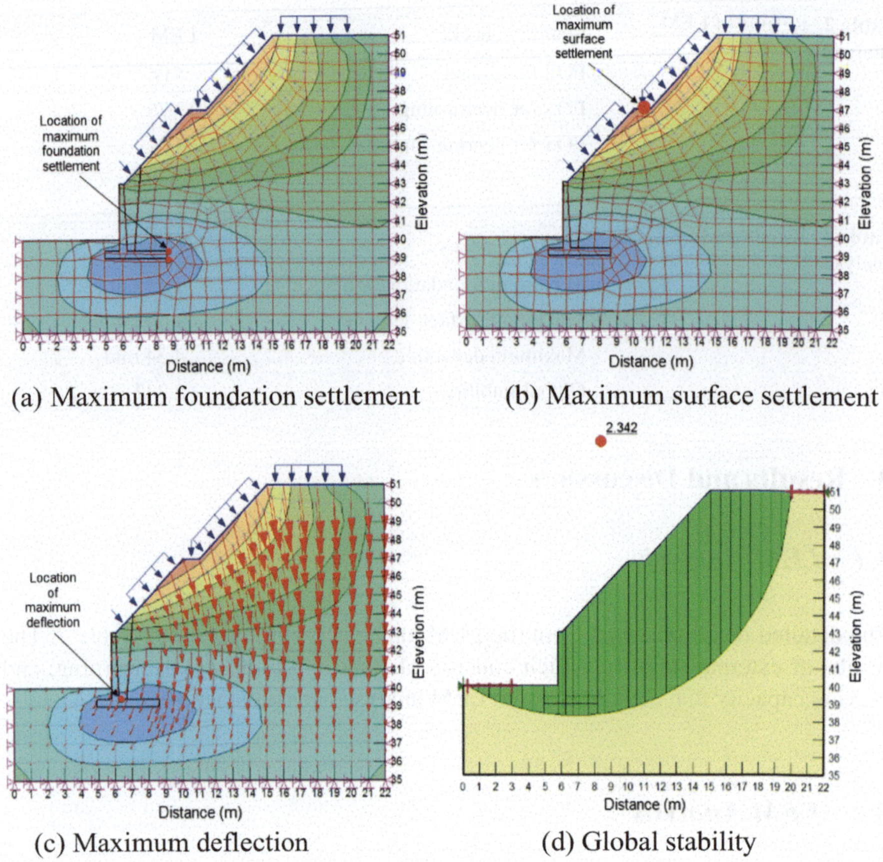

Fig. 3 Results of FEM analysis for **a** maximum foundation settlement, **b** maximum surface settlement, **c** maximum deflection, and **d** global stability

4 Conclusions

The comprehensive design of concrete cantilever retaining wall was produced from the combination of LEM and FEM methods which contain external and internal stability. The external stability analysis consists of FOS for sliding, overturning, and bearing capacity. Meanwhile, the internal stability analysis involves maximum foundation settlement, maximum surface settlement, maximum deflection, and global stability. The finite element method is very useful as it can provide much additional information, such as stress distribution and displacement that is difficult to obtain from model tests.

As conclusion, this study can address the shortcomings in design of concrete cantilever retaining walls that do not take into account the foundation settlement, surface settlement, deflection, and global stability in design calculations. It can help the consulting engineers produce a comprehensive design, and the stability of concrete cantilever retaining wall can be enhanced.

Acknowledgements The authors would like to acknowledge UiTM Pahang for funding ICSDEMS 2019 fee.

References

1. Dhamdhere, D. R., Rathi, V. R., & Kolase, P. K. (2018). Design and analysis of retaining wall. *International Journal of Management, Technology and Engineering, 8,* 1246–1263.
2. Vijayakumar, S., Subhash, A. B., Sumant, T. P., et al. (2015). Case study on static stability analysis of retaining wall at Dewarwadi. *International Journal of Innovative Research in Science, Engineering and Technology, 4,* 3071–3077.
3. Yadav, P. A., Padade, A. H., Dahale, P. P., et al. (2018). Analytical and experimental analysis of retaining wall in static and seismic conditions: A review. *International Journal of Civil Engineering and Technology, 9,* 522–530.
4. Tasildar, B. S. (2015). Stability of retaining wall under seismic load: A review. *International Journal of Engineering Sciences & Research Technology, 4,* 324–326.
5. Cernica, J. N. (1995). *Geotechnical engineering: Soil mechanics.* United States: Wiley.
6. Punde, G. V., Auti, A. S., Yendhe, R. R., et al. (2018). Design of retaining wall. *International Journal of Advance Engineering and Research Development, 5,* 1–7.
7. Patil SM, Wagh KS (2010) Reduction in construction material: Effect of the provision of the loft behind the cantilever retaining wall. In *Indian Geotechnical Conference* (pp. 227–230)
8. Padhye, R. D., Ullagaddi, P. B. (2011). Analysis of retaining wall with pressure relief shelf by coulomb's method. In *Proceedings of Indian Geotechnical Conference* (pp. 671–673)
9. Aswin, L. (2018). Lesson learned from retaining wall failures: A geotechnical disaster. In *MATEC Web of Conferences* (vol. 229, p. 03014).
10. Chetan, S., & Vijay, B. (2014). Evaluation of the effect of lateral soil pressure on cantilever retaining wall with soil type variation. *Journal of Mechanical and Civil Engineering, 11,* 36–42.
11. Asmaa, G. S., & Heba, A. A. (2014). The effective contribution of software applications in various disciplines of civil engineering. *International Journal of Civil Engineering and Technology, 5,* 316–333.
12. Huy, T. P., Htet, Z. O., & Cheng, J. (2013). Stability of slope and seepage analysis in earth dam using numerical finite element model. *Study of Civil Engineering and Architecture (SCEA), 2,* 104–108.

Use of Waste Low-Density Polyethylene (LDPE) as Bitumen Modifier in Asphalt Concrete-Binder Course (AC-BC) Mix

Sukobar, Machsus Machsus, Rachmad Basuki, Amalia Firdaus Mawardi, and Moh. Firli Firdausi

Abstract The purpose of this study is to investigate the possibility of using low-density polyethylene (LDPE) plastic waste as a bitumen modifier for asphalt concrete-binder course (AC-BC) mixtures. The influence of LDPE obtained by various mixing temperatures and LDPE content on the Marshall properties on road pavement was investigated. LDPE mixing process in this study uses the dry method. It means that LDPE plastic material is directly mixed into aggregate; when the aggregate is heated, then the mixture of aggregate and LDPE is added to hot asphalt. In this research, asphalt with penetration of 60/70 is used. Its optimum asphalt content (OAC) is 5.65%, while LDPE added are 2, 4, and 6% of the asphalt by weight. Each variation of LDPE contents was carried out making 5 specimens. The results showed that the addition of LDPE plastic waste into the AC-BC mixture had a positive effect on Marshall properties on road pavement. The addition of LDPE plastic waste has been able to increase the value of stability and Marshall quotient (MQ). Besides that, the values of Marshall properties such as voids in mix (VIM), voids in mineral aggregates (VMA), voids filled with asphalt (VFA), and flow also still meet Indonesian National Standard (SNI).

Keywords Plastic waste · LDPE · AC-BC · Marshall properties

1 Introduction

Recently, waste is a problem that is being faced by all countries in the world, including Indonesia. Indonesia is in the second position in the world with the amount of plastic waste in the sea which reaches 1.29 million tons per year. Plastic waste has a bad impact on the environment, because it is difficult to be decomposed by the soil naturally. Trials on the use of plastic waste for road construction have been carried out by the Ministry of Public Works and Housing (PUPR). Construction of a kilometer-long highway is estimated to require 2–5 tons of plastic waste [1].

Sukobar (✉) · M. Machsus · R. Basuki · A. F. Mawardi · Moh. Firli Firdausi
Department of Civil Infrastructure Engineering, Institut Teknologi Sepuluh Nopember, Surabaya, Indonesia
e-mail: kobarabiyi@gmail.com

S. S. Emamian et al. (eds.), *ICSDEMS 2019*,
https://doi.org/10.1007/978-981-15-3765-3_18

Low-density polyethylene (LDPE) is a type of plastic waste that has the potential to be used as an added material in flexible pavement. The characteristics of this plastic waste which are made from petroleum have a low density of 0.91–0.94 g/cm^3, and a melting point of 105–115 °C. In this study, an investigation was made about the possibility of using LDPE-type plastic waste as a bitumen modifier for asphalt concrete-binder course (AC-BC) mixtures [2].

2 Materials and Methods

This research uses the following materials: waste plastic bags, asphalt, Portland cement, and aggregate combined. Figure 1 shows the composition of the combined aggregate used in the AC-BC mixture in accordance with the SNI Specifications consisting of: 13% coarse aggregate (10–20 mm), 20% coarse aggregate (10–10 mm), medium aggregate 20% (5–10 mm), 45% fine aggregate (0–5 mm), and 2% filler.

The addition of LDPE plastic waste to the AC-BC mixture in this study uses the dry method. This means that LDPE particles are added to the hot aggregate, and stirred until the LDPE particles cover the aggregate. Furthermore, hot asphalt is poured into the aggregate, which has been mixed with LDPE plastic waste. This method was chosen because it is easier and cheaper than the wet method [2, 3].

The composition of asphalt content used for calculation of optimum asphalt levels (OAC) is 5.65%. The composition was calculated based on an aggregate weight of 1200 g, and an OAC value of 5.65% was obtained. Besides, the proportions of the addition of LDPE are 2, 4, and 6% to the weight of asphalt content. These additional particles of plastic waste can reduce the total weight of asphalt in the AC-BC mixture.

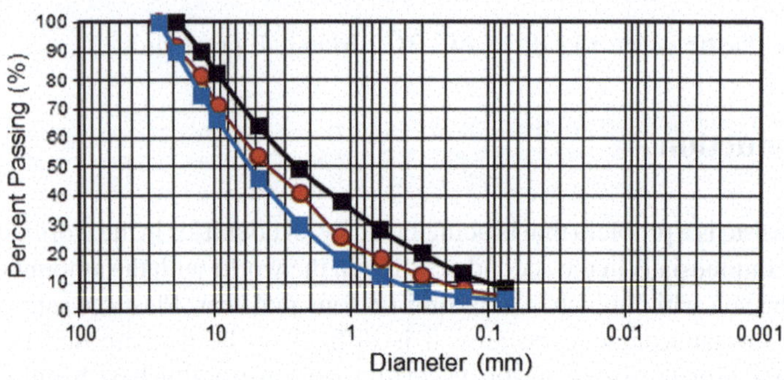

Fig. 1 Aggregate blending

3 Result and Discussions

The effect of using LDPE obtained with various levels of asphalt changes the Marshall properties values. Marshall properties observed and tested in this study consisted of: density, void in mix (VIM), void in mineral aggregate (VMA), void filled asphalt (VFA), Marshall flow, Marshall stability, and Marshall quotient. Marshall test results on the AC-BC mixture with the addition of LDPE can be explained as follows.

Figure 2 shows that the density values obtained between the AC-BC mixture without and with the addition of plastics are almost the same, with a maximum difference of 0.01 g/cm^3. This means that the addition of LDPE plastic waste does not significantly influence the density value of the AC-BC mixture, because the proportion is only up to 6% of the optimum asphalt content (OAC). In other words, the difference in density values in the variation of the addition of plastic is more influenced by other factors [4].

VIM values are obtained in the AC-BC mixture without and with the addition of plastic coinciding with the lower bound value in the provisions of the Indonesian National Standard (SNI) as shown in Fig. 3. All variations of the mixture also met the requirements, except AC-BC mixture with the addition of 6% plastic. The VIM value indicates that the percentage of cavities in the AC-BC mixture is indeed relatively small or not porous. However, the characteristics of mixtures with VIM values lower than 3% have the potential to cause plastic grooves and swelling, even though they are waterproof and do not easily crack early [4, 5].

Figure 4 shows that the VMA value obtained in the AC-BC mixture without and with the addition of plastic is above the minimum boundary, which is 14%. A minimum VMA limit is needed to prevent the occurrence of asphalt shortages in the AC-BC mixture, because it causes the granules in the mixture to break easily and crack. VMA value that meets SNI standards affects the durability of road pavement,

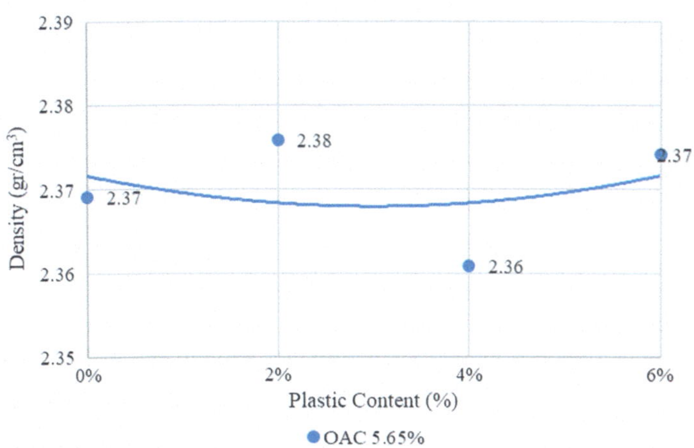

Fig. 2 Relationship of density and plastic contents

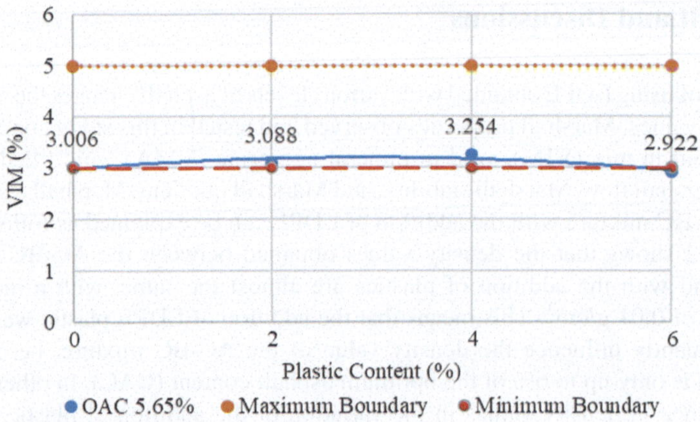

Fig. 3 Relationship of VIM and plastic contents

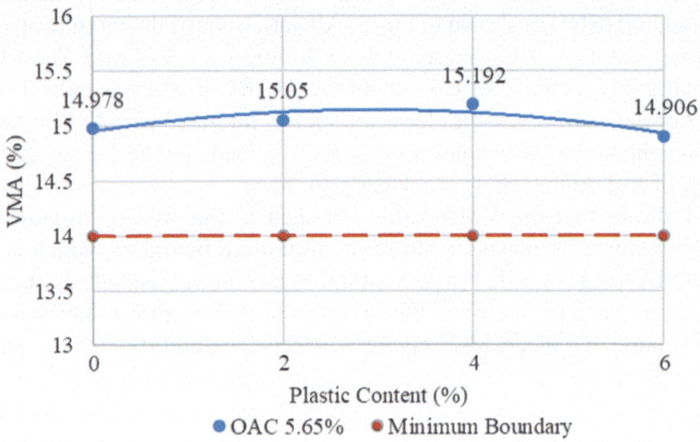

Fig. 4 Relationship of VMA and plastic contents

so the level of road service can be maintained according to the pavement age. The maximum difference between the AC-BC mixture without and with the addition of plastic is only 0.214%. The maximum difference occurs in the addition of 4% plastic content. The graphic illustration in Fig. 4 shows that the addition of LDPE plastic waste can increase the VAM value at the plastic content of 2–4%, but at a plastic level of 6% there is a decrease [4].

The VFA value obtained in the AC-BC mixture without and with the addition of plastic is above the minimum boundary (65%) as shown in Fig. 5. The maximum difference between the AC-BC mixtures without and with the addition of plastic is 1436%. The maximum difference in VMA value occurs in the addition of 4% plastic content. The addition of LDPE plastic levels can reduce the VAM value in the plastic

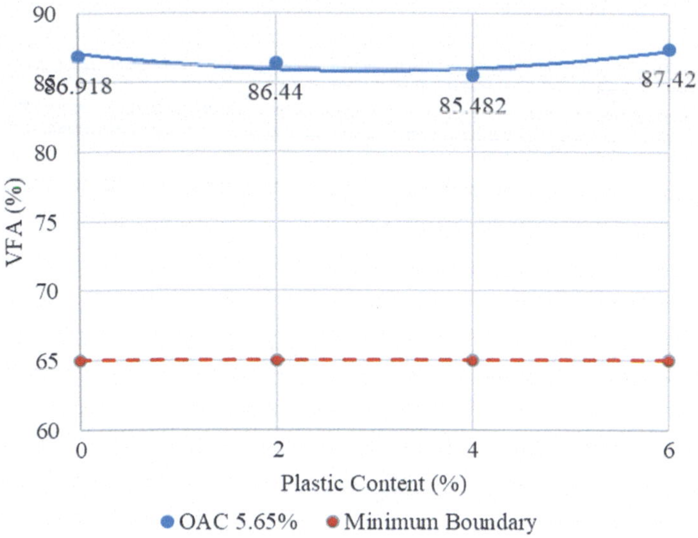

Fig. 5 Relationship of VFA and plastic contents

content of 2–4%, but at a plastic level of 6% there is an increase. VFA functions to cover aggregate grains in the AC-BC mixture. A minimum VFA of 65% is important to prevent road pavement from being worn out. VFA values that meet the standards can prevent the mixture from rutting easily against heavy traffic loads [6].

The graph in Fig. 6 shows that the addition of plastic content to the AC-BC mixture can increase the stability value with a very high increase above the minimum boundary. The stability value in the AC-BC mixture without the addition of plastic

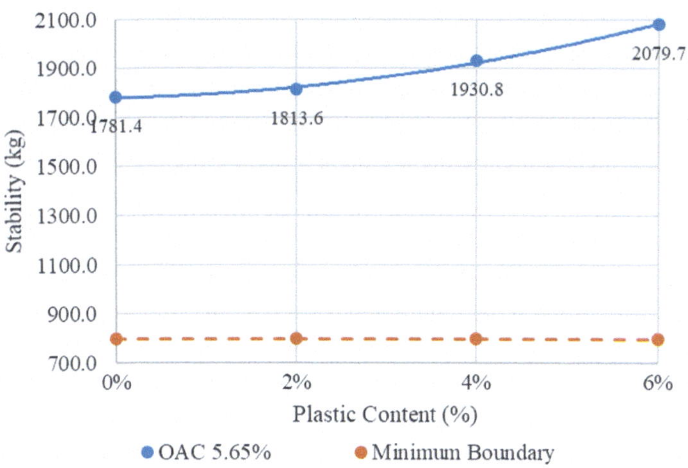

Fig. 6 Relationship of stability and plastic contents

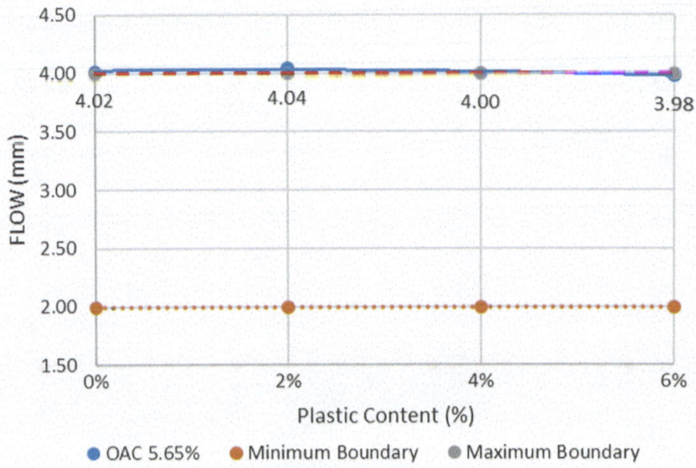

Fig. 7 Relationship of flow and plastic contents

was 1781.4 kg. The stability value was increased to 1813.6 kg with the addition of 2% plastic content, and increased again to 1930.8 kg for 4% plastic content. The highest stability value of 2079.7 kg occurred in the addition of plastic content by 6%. This shows that the addition of LDPE plastic waste causes locking between aggregate particles and the binding capacity of modified asphalt to aggregate becomes stronger [3, 7].

From Fig. 7, it is known that the flow value obtained in the AC-BC mixture without and with the addition of plastic coincides with the maximum boundary value in accordance with the provisions of the Indonesian National Standard (SNI). All variations of the mixture also meet the requirements between the minimum and maximum boundaries. The flow value shows that the AB-BC mixture is able to adapt to changes in shape due to loading, so that the pavement does not crack easily. But the amount of flow is also limited to prevent waves and grooves in the pavement, so that the pavement provides comfort and safety of traffic [1, 5, 7].

Adding plastic content to the AC-BC mixture can increase the MQ value very high above the minimum boundary, as shown in Fig. 8. The MQ value in the AC-BC mixture without the addition of LDPE plastic waste was 430.1 kg/mm. The MQ value decreased slightly to 430 kg/mm in addition of 2% plastic content, then increased again to 462 kg/mm in addition of 4% plastic content, and increased again to 500.5 kg/mm in addition plastic content by 6%. All variations of the AC-BC mixture have also fulfilled the requirements because the MQ values were obtained above 250 kg/mm [8, 4].

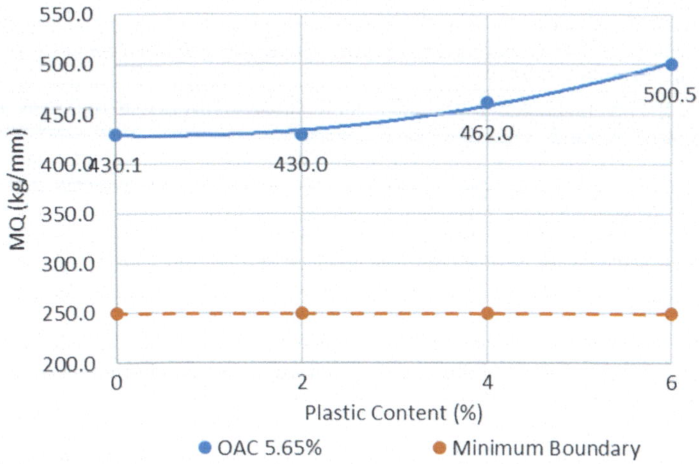

Fig. 8 Relationship of MQ and plastic contents

4 Conclusions

Based on the literature review, analysis, and discussion, it is concluded that [1]: The results showed that the addition of LDPE plastic waste into the AC-BC mixture had a positive effect on Marshall properties on road pavement. The addition of LDPE plastic waste has been able to increase the value of stability and Marshall quotient (MQ). Besides that, the values of Marshall properties such as voids in mix (VIM), voids in mineral aggregates (VMA), voids filled with asphalt (VFA), and flow also still meet the Indonesian National Standard (SNI) [2]. The results of this study are believed to be able to contribute positively to the development of sustainable pavement material technology for road infrastructure works. The composition of combined aggregate from the AC-BC mix could be formulated as follows: 13% of coarse aggregate (10–20 mm), 20% of coarse aggregate (10–10 mm), 20% of medium aggregate (5–10 mm), 45% of fine aggregate (0–5 mm), and 2% filler.

Acknowledgements The authors would like to express their gratitude to the reviewers for the constructive remarks. We also would like to thank ITS Surabaya for the financial support provided through Research Grant.

References

1. Purnamasari, P. E., & Suryaman, F. (2010). Sebagai Bahan Tambah Terhadap Karakteristik Lapis Aspal Beton (Laston). *Konf. Nas. Tek. Sipil 4 (KoNTekS 4)*, *4*(4), 2–3.
2. Gawande, A. (2016). *A review utilization of waste plastic in asphalting of roads* (Vol. 2, pp. 147–157, 2016.

3. Sultana, K. S. B., & Prasad, S. A. (2012). Utilization of waste plastic as a strength modifier in surface course of flexible and rigid pavements. *International Journal Engineering Research Application, 2*(4), 1185–1191.
4. Aschuri, I., Woodward, D., & Woodside, A. (2005). *The use of waste materials to modify binder in asphalt concrete* (pp. 1–11).
5. Eme, D. B., Nwofor, T. C., & Umukoro, E. S. (2015). Stability of asphalt cement concrete produced from waste plastics as replacement for aggregate. *International Journal Civil Engineering Technology, 6*(5), 976–6308.
6. Nkanga, U. J., Joseph, J. A., Adams, F. V., & Uche, O. U. (2017). Characterization of Bitumen/Plastic blends for flexible pavement application. *Procedia Manufacture, 7,* 490–496.
7. Yasanthi, R. G. N., Rengarasu, T. M., & Bandara, W. M. K. R. T. W. (2016). *Variation of Marshall properties of hot mix asphalt concrete with waste materials* (pp. 75–81).
8. Vasudevan, R., Ramalinga Chandra Sekar, A., Sundarakannan, B., & Velkennedy, R. (2012). A technique to dispose waste plastics in an ecofriendly way—Application in construction of flexible pavements. *Construction Building. Materials, 28*(1), 311–320.

Comparison of Energy Efficiency and Thermal Comfort Technology: Case Studies of Tsinghua Eco Studio & S11 House

Tan Jia Moon, Nangkula Utaberta, Chia Lin Lin, and Nayeem Asif

Abstract In the olden days, the passive design was taken for granted in the traditional architecture, as the technology was not so advanced. In modern days, the advance of technology has provided greater comfort with the cost of energy consumption. By foreseeing the future, the fellow profession has discussed energy efficiency, particularly in the building industry, to reduce the rate of resource depletion. Besides that, energy efficiency is also directly affecting the thermal comfort of the occupants within the building. Buildings that were constructed within the tropical region can incorporate their advantage of small temperature difference into their building design and technology. Therefore, this paper is discussing the comparison between the differences in green innovation from two buildings, which are located in two countries with a similar climate.

Keywords Energy efficiency · Thermal comfort · Passive design · Tropical region · Green building

1 Introduction

A building consumes massive energy when it is occupied. As energy consumption costs a huge amount of resources, it is contributing to the energy depletion of the world. Therefore, green building has become popular to reduce the energy consumption of a building. Energy efficiency is one of the key issues that is highly discussed

T. J. Moon · C. L. Lin
School of Architecture and Built Environment, FETBE, UCSI University, Cheras, Malaysia

N. Utaberta (✉)
Halal Products Research Institute, Universiti Putra Malaysia, Serdang, Selangor 43400, Malaysia
e-mail: nangkulautaberta@gmail.com

Faculty of Built Environment, Linton University College, Mantin, Negeri Sembilan, Malaysia

N. Asif
Kulliyah of Architecture and Environmental Design, International Islamic University Malaysia, Gombak, Malaysia

S. S. Emamian et al. (eds.), *ICSDEMS 2019*,
https://doi.org/10.1007/978-981-15-3765-3_19

in the industry of green building. Therefore, this paper takes a step further to discuss the difference between green innovations from two buildings located in the countries with a similar climate.

2 Literature Review

This paper is studying the innovative technologies that can contribute to energy efficiency and thermal comfort. The first case study is S11 House. It is Malaysia's first GBI Platinum-rated house. It is located at the older suburban area in Petaling Jaya. It is owned and designed by Dr. Tan Loke Mun. The concept of the house is drawn from the idea of space under the shade of large tree canopy. The green tropical house is surrounded by five significant trees on the site. These trees are three old frangipanis, a largssse star fruit tree and a coconut palm tree [1]. While the second case study is Tsinghua Eco Studio. It is the first demonstrating and experimental Nearly Zero Energy Building of the Gui'an Innovation Park in southwest China. It is a project derived from the joint program by Tsinghua Holdings Human Settlement Construction Group (THHSCG) and BRE (Building Research Establishment) from the UK. The idea is to build a sustainable building that fulfills the requirement of the Building Research Establishment Environmental Assessment Method (BREEAM). The studio is connected with the constant ecological sponge city landscape, providing beautiful scenery for the guests. The two floors above ground can serve as a large exhibition space, while one single underground floor can act as facilities and storage area. The modular double-skin façade system is a unique integration of vernacular rattan-weaving craftsmanship with industrial prefabrication technology, highly expressing site-specific characteristics while promoting local traditional industry and economy [3].

3 Research Methods

This study adopts a qualitative strategy to acquire secondary data and analyzes them through descriptive analysis. Two study cases have been selected based on their relevancy on the current topic. Findings from the study cases are compared to derive a conclusion on how efficiently the cases can resolve the issue in hand. The final results are discussed at the end of this paper.

4 Analysis and Results

4.1 Case Study 1: S11 House

The first case study is S11 House. For energy efficiency, lightweight concrete blocks of the house are insulated with heat reflecting paint applied. This helps to lower down the heat absorption from radiation. The house has been built with north–south orientation for its openings to avoid direct heat gain from the sun yet still receive some daylight into the building, while the east and west walls have less opening. Large tree-like canopy roof is constructed of lightweight recyclable profiled steel metal sheets coated in a light, off-white color to minimize heat absorption. The indoor environmental quality of S11 is related to the method of bricklaying. Bricks are laid with hollow spaces between one another to allow for ventilation into the building. This also enriched the user experience through the penetration of light into space [1].

Open spaces were allowed for landscaping which are able to reduce the heat island effect of the area. The local materials, like bricks and hardwood from the debris of the demolition, were then reused for the new construction. In the house, there is a total of 15 specially designed wind turbines with a steel framed glazed pyramid which provides stack effect ventilation and light in the house [2]. The turbines are driven by wind and convection of air within the glass pyramid, which is heated up as a result of the greenhouse effect [1].

4.2 Case Study 2: THE-Studio

The development of multi-system integrated design strategy is aimed to minimize the negative impacts on the local ecological system by minimizing the construction period while maximizing the indoor comfort and energy efficiency. The multi-system integration includes three levels: the parallel construction systems, the integration of vernacular culture with sustainable technology, and the building information model (BIM) platform.

The building consists of timber frameworks, prefab light-steel modules, sustainable services, and modular façade systems. All these systems are implemented in parallel to accelerate the whole construction process, thus reducing negative interference to the site. For example, the timber frameworks are assembled in the large-space exhibition hall, while the prefab light-steel modules are used to shape functional space at both sides. All these structural components and joints are produced off-site and swiftly assembled on site, leading to a significant saving of time and energy.

Sustainable service systems, including air through tunnel, biomass heating, photovoltaic thermal (PVT) system, rainwater collecting, control, and monitoring, are mostly plugged into the cavity of façades, which can save installation time, increase indoor flexibility, and provide potential space for additional equipment in future experiments [5].

BIM is adopted as integrating platform throughout building's whole life span, especially for coordination of construction, operation, and maintenance phases. A series of measurements and monitoring for thermal conditions, humidity, ventilation, and luminous environments were carried out and further analyzed. The results verified that the detailed design methods and building technologies were effective and appropriate during construction and operation stages. This information could serve as a reference for other sustainable buildings in southwest China.

During the design stage, the architects still place their priority in terms of layout and space planning, shape, and material selection to make the building adopt the local climate. These are to ensure that the building can maximize the indoor environmental comfort by obtaining more natural light, ventilation, and sunlight control in a simple manner.

The exhibition hall has a ventilating skylight to act as a chimney to promote ventilation; at the same time, it also provides colorful light into the interior, creating some visual elements inside the building. As for the second floor, four different kinds of the rattan-weaving pattern are arranged according to the information local solar radiation and wind pressure through software simulation. It aims to integrate environmental performance, structural stability, and material durability [4].

The building also uses underground air through the tunnel as the passive air-conditioning system, which can largely reduce energy consumption. The vertical venting ducts are integrated into the air cavity of the double-skin façade and distributed into the major working spaces, providing fresh air while creating indoor comfort.

The double-skin facades consist of double-glazed façade on the first floor and rattan-weaving double skin on the second floor. The combination of two façade treatments are the climate responsive design elements for this building. Through different operative modes of vents and windows, the double-glazed façade can adapt the seasonal and daily changes of the outdoor environment and reach the expected thermal and ventilation performance.

Besides that, the building encourages the use of renewable materials such as timber, steel, and wheat-straw board and promotes the employment of vernacular materials and craftsmanship such as traditional rattan-weaving, blue-stone floor paving, and rubble wall masonry, which can both decrease the carbon footprint during the building's life span and create a unique architectural expression.

5 Discussion and Conclusion

Despite the fact that two buildings are in different categories, they are brought into the comparison due to the similarity in size and climate—the two buildings are located in the tropical and humid subtropical regions. Therefore, the concepts of their green design innovation are alike.

Both green design features are designed with components to create a stack effect by inducing convection of air inside the components. In my opinion, however, the design innovation for S11 House is not fully integrated with the building design. It looks like an add-on to the building. In the interior of S11, it has been shown that the 3 pipes are extended from the basement penetrating through the first floor to the roof, while the design innovation for THE-Studio is not a separated entity, but part of the building.

References

1. ArchDaily. (2013). S11 House/ ArchiCentre.
2. McDonald, A. (2015). How Malaysia's S11 house brings modern style to Petaling Jaya.
3. Ramli, N. H. (2012). Re-adaptation of Malay house thermal comfort design elements into modern building elements—Case study of selangor traditional Malay house and low energy building in Malaysia. *Iranica Journal of Energy and Environment, 3,* 19–23. https://doi.org/10.5829/idosi.ijee.2012.03.05.04.
4. SUP Atelier. (2015). Song und partner atelier.
5. The Architecture MasterPrize. (2019). The-studio (Tsinghua eco studio).

Sustainability in Architectural Conservation of Heritage Building: A Qualitative Approach

Yeo Shu Han, Nangkula Utaberta, Nayeem Asif, and Norizan Daud

Abstract While there are many heritage buildings with over 50 years of age in Malaysia, many are then left abandoned to decay. It is clear that given proper planning and consideration for conserving these existing structures, we are presented with opportunities for adaptive reuse of buildings well designed to respond to the environment through passive means, without the need to demolish or clearing the existing green sites for new constructions. However, conservation of heritage architecture is not often discussed as part of the sustainable movement of architecture. This paper will discuss the process and aspects of architectural conservation in Malaysia and its relevance to the green building guideline.

Keywords Conservation · Heritage · Architecture · Sustainability · Bangunan Sulaiman

1 Introduction

The idea of sustainable development started surfacing in the mid-1900s as scientists began shedding lights on the pressing issue of environmental pollution crisis. With that, the concept of green building and green building rating tools came about with UK pioneering through BREEAM.

At the mention of sustainable architecture, most will immediately relate it to building energy performance and building design strategies: passive strategy through

Y. S. Han · N. Daud
School of Architecture and Built Environment, FETBE, UCSI University, Cheras, Malaysia

N. Utaberta (✉)
Halal Products Research Institute, Universiti Putra Malaysia, Seri Kembangan, Selangor 43400, Malaysia
e-mail: nangkulautaberta@gmail.com

Faculty of Built Environment, Linton University College, Mantin, Negeri Sembilan, Malaysia

N. Asif
Kulliyah of Architecture and Environmental Design, International Islamic University Malaysia, Gombak, Malaysia

S. S. Emamian et al. (eds.), *ICSDEMS 2019*,
https://doi.org/10.1007/978-981-15-3765-3_20

means of vernacular strategy, active technology, and innovation that emphasizes on renewable energy and energy efficiency. However, design is only one out of many aspects encompassed in the concept. Sustainable architecture takes into account all aspects of a building's life cycle from design, construction, operation, and maintenance to demolition, and analyses the process at every stage on its impact on the environment. It is about the selection of materials, the method materials sourced, the logistical aspect of transporting the materials to site, and finally, how much carbon footprint is imposed throughout the life cycle of the building.

> The greenest building are the ones that are already built [1].

In this context, Malaysia being the colony of many countries prior to its independence houses many heritage buildings with other countries' influences, many of which are left derelict and unmaintained. Given proper maintenance and planning of use to the size of existing buildings, the adaptive reuse of heritage buildings would significantly reduce the impact on the environment by eliminating the process of demolition and the planning and clearing of sites for new buildings. This paper will focus its discussion on the sustainability of architectural conservation throughout the process of conservation in the stage of construction works in Malaysia.

2 Literature Review

2.1 Architectural Conservation

Conserving national heritage assets and places retains and provides sense belonging to the people in a built environment. Sense of belonging, shared heritage, and historical background in a diverse cultural society such as Malaysia are crucial in bringing citizens together which contribute to nation building [2]. Each nation has its 'historic buildings' that represent its past. Historic buildings are directly linked to nationalism, through their roles in building and reinforcing national identity. Despite their physical existence, historic buildings are 'created' rather than given: they must be constructed as 'historic', through processes of choice and the attachment of significance [3].

One of the principal functions of heritage interpretation is to enhance the visitor's sense of place and place identity. If this is to occur, the interpretation should be planned and designed with that outcome in mind [4]. The real heritage value is not just in the physical structure, but in the stories that are part of that building and its surroundings. Heritage building conservation is no exception, which essentially comprises the physical evidence of our environment that symbolizes the tangible cultural identity and heritage of the nation. In the case of Malaysia, it is a means of affirming our national heritage and promoting solidarity, thus providing the means of satisfying a wide variety of aspirations. The Malaysian government initiates relevant legislations and enforcements as they play important roles in conservation of heritage buildings to safeguard the spirit and identity of the nation.

Linkage to past glories is often connected to using historic buildings as national rallying symbols in the face of external threats. Wars and invasions have seen remarkable changes in policy, as historic buildings of the previous opposed regimes or internal groups become adopted as national symbols [3]. Hence, architectural conservation is an important effort to sustain the historical narration of the city's existence.

2.2 Life Cycle Assessment of Heritage Building

The true concept of conservation is preserving the authenticity of the heritage based on the original or historical evidence. Authenticity is a process or desire to reveal the true nature of an object. The United Kingdom Guidance for Practice defines conservation as the means by which the true nature of an object is preserved. The true nature of an object includes evidence of its origins, its original construction, and the materials by which it is composed and information as to the technology used in manufacturing [5].

Heritage presents itself in both tangible and non-tangible forms as a narrative of a nation. In the context of architecture, heritage building is a monumental representation of a country in its identity, history, and culture—through style of architecture, as a form of memory and reminder of historical events that formed the nation that we are today. Conservation can be seen as an effort to prolong the operational timeline of a building, as well as an effort to maintain nation's identity.

Building and structure gazettes as heritage are those that hold the value of history or culture for more than 50 years. Given such, when technology and innovation were not as advanced before the 1960s, buildings would rely solely on passive means of design to achieve optimum indoor environmental quality of lighting and temperature condition [1].

2.3 Case Study: Bangunan Sulaiman

Bangunan Sulaiman is one of the last three in the series of heritage buildings which were constructed in Kuala Lumpur by the British. It is significant as an important architectural heritage to our nation due to its location in the context with Kuala Lumpur Railway Station, the current Majestic Hotel, Kuala Lumpur Railway Office, Masjid Negara, Merdeka Square, and Stadium Merdeka, forming the prominent heritage strip in Kuala Lumpur City Centre.

The building started its operation in 1930 as the head office to Federated Malay States Railways, and it currently functions as the office for Asian International Arbitration Centre (AIAC) since its conservation in 2011 [6]. The building adopts the architecture language of art deco with implementation of simple geometrical forms of ornaments on minimal areas of its façade expressed through clean lines and rectangular forms.

> In addition to the original building which will be refurbished, the project involves the development of new two-story car parking facility with a pavilion housing training facility on the existing open car park space to the west of the original building. To the south, at the rear end of the original building will be the two new structures – a single story open cafeteria and a five-story block which will hold lifts and a staircase [7].

3 Interview with Conservation Architect of Bangunan Sulaiman

An interview was held on 19 September 2018, at the interviewee's office in PJU with the objective of understanding the process and challenges faced in the conservation of Bangunan Sulaiman. The researchers presented the following questions to Ar David Cheah, the architect involved in Bangunan Sulaiman's conservation:

Research questions	Significant notes from the answers
What is the role of the architect in the conservation of Bangunan Sulaiman?	• Visitation and study of selected heritage building • Restoration of building to original drawings from national archive while ensuring interior planning is suitable for current usage • Build a car park block next to Bangunan Sulaiman to increase the functionality of building
What are the problems faced in the conservation and the solutions taken?	• Constraint due to limited budget for high-cost building restoration • No physical contact allowed between heritage and new structures, exterior to remain unchanged • New structure required by DBKL to respond to heritage building while not outshining to respect its authenticity • Worker's lack of knowledge on specialized construction technique for specified building materials to match the original building that are often acquired from overseas
What is the relationship and importance between heritage conservation and the nation? What is the role of heritage architecture towards nation building in Malaysia	• Heritage building is of historical value to our country, thus the need for conservation • As aspect of tourism to boost the country's economy • Architecture as narration to represent our identity • Modern examples of architect contributing to nation building in Malaysia are Hijjas Kasturi and Ken Yeang

(continued)

(continued)

Research questions	Significant notes from the answers
What are the suggestions to strengthen the awareness and pride in heritage building among our nation?	• Educate the young generation, students, as well as public through adaptive reuse and experiential activities • Architects should design better or flexible space and function for the heritage buildings in order to increase the public engagement and awareness to allow the users or future users to appreciate the heritage building • Heritage bodies and authority to align and strengthen the conservation policy

4 Methodology

This study adopts qualitative strategy to acquire secondary data and analyses them through descriptive analysis. Research through journals and literature reviews is selected via online and Chen Voon Fee Resource Centre at Badan Warisan Malaysia using keywords of heritage and conservation. Text analysis is then applied to examine contents relevant to the architectural conservation in Kuala Lumpur and its relation to sustainability towards environment and the nation. Bangunan Sulaiman is selected as the case study to analyse the sustainability throughout its approach of architectural conservation. Data on Bangunan Sulaiman is collected through site observation through visitation and interview with the conservation architect of Bangunan Sulaiman, AR. David Cheah. The final results are discussed at the end of this paper.

5 Results and Discussion

While there are many heritage buildings with over 50 years of age in Malaysia, many are then left abandoned to decay. It is clear that given proper planning and consideration for conserving these existing structures, we are presented with opportunities for adaptive reuse of buildings well designed to respond to the environment through passive means, without the need to demolish or clearing of existing green sites for new construction. Unfortunately, as presented by the reality, there is a lack of attention given to the process of maintenance and restoration work, and hence, these buildings are left to be demolished when the structure decays beyond its function.

Green Building Index, while it should not be the limiting guideline to sustainable construction, is undeniably a benchmark that professionals in the construction industry refer to measure the sustainability of their building projects. As we take a look at the categories in Green Building Index (GBI), while there are categories listed for existing buildings, the breakdown is without consideration to the intricate requirement expected for the gazetting of heritage buildings. Putting aside the intangible narratives in the expectation of society and culture that ties a building to any historical events, for a building to be gazette as heritage, it must retain to its true nature. It must retain to good design or aesthetical characteristics that is the architectural elements and features of original design, construction, and material that represent its respective historical event.

By making a comparison between the requirements of Green Building Index rating tool and National Heritage Department, the contradiction which would not allow the eligibility of heritage buildings is not allowed to be rated green or sustainable in Malaysia as shown. This indicates that there is a gap in the green building guideline that could be further revisited to accommodate to heritage buildings.

5.1 *Building Design and Orientation*

As society advanced with technological innovation, we began relying on mechanical and technology means of solution. There are usually minor issues of overall building heat gain in heritage buildings. Issues are overcome in overall site planning and orientation of buildings to ensure minimum building surface and openings are exposed to the east–west axis of direct sun path.

Taking case study of Bangunan Sulaiman as an example, the building shows that Bangunan Sulaiman sits in a longitudinal position along the east–west sun path, thus receiving minimal sun heat gain. The plans also indicate further buffer provided by corridor surrounding the perimeter of the building. As wind travels in the south-east direction, cross-ventilation is achieved at its optimum. However, due to the function of the building for arbitration use, privacy of spaces is required and concealed.

SM5 in non-residential existing building category of GBI rating indicates that parking capacity should be achieved for sustainable site management. Site of such heritage buildings is often located in urban settings, as colonial countries develop their administrative centres and township. Heritage requirements indicate that there should be no extension to the existing structure as it differs it from its true form but additional structure may be constructed within the site of heritage building as long as it does not touch the existing structure. In cases such as Malacca where the heritage town is already congested, there is no land available nearby the site and thus, it would require a change in town planning and system to achieve parking facility to accommodate for the building function and provide proper public transportation linkage.

5.2 Building Material and Maintenance

Extracted from the interview with Ar David Cheah, it is understood that the statutory body for heritage in Malaysia requires the restoration works of the building façade to be done entirely with its original materials. More often, similar in the case of Bangunan Sulaiman, the specified materials are not available in our local context, hence requiring sourcing and purchasing from a foreign country. In this term, it has significantly reduced the sustainability of the building in the aspect of economy through the increase in overall project cost of obtaining material and impact on the environment by imposing of carbon footprint through the need of logistical solution. Labour cost will also increase as many of the restorations consist of ornaments and details that are beyond conventional construction, thus requiring skilled workers or specialists.

Besides that, as technology was not as advanced during the time of construction for buildings listed as heritage, innovation of building materials was similarly quite limited in choice. With that, materials found in the building may not be the most efficient when it comes to durability. Because of this, by emphasizing the use of exact same material and not allowing replica, it will require frequent maintenance and replacement works to the building with similar materials and hence to reduce the overall sustainability of the building in the long run.

As heritage body has strict requirements for the building façade to be entirely replicated in terms of form and materials, it is not possible to achieve MR1, MR2, and MR6.

5.3 Cultural Sustainability and Nation Building

Efforts of architectural conservation are important and significant as heritage buildings are the physical symbolism and reminder of our nation's common identity. The heritage buildings are monuments left in time of representation of our history.

While awareness of importance of architectural conservation is given effort and emphasis, the lack of funding system by the government in supporting the conservation projects and lack of skilled workers with knowledge of applying materials specified as the building's original form in Malaysia become an issue.

Besides that, heritage buildings conserved in city of Kuala Lumpur are mostly in terms of their external façade only. The lack of interaction of the people with the buildings lacks opportunity to create memory and connection to bond the people to embrace and appreciate the buildings as representation of nation's identity. Therefore, conservation of heritage buildings with proper integration and programme planning to engage people in using the spaces provided allows better appreciation of people to the buildings due to personal connection from memories created.

Once the connection is established between the people and architecture heritage, a common identity can be visualized in aid of nation building. Thus, the people will garner more will and interest to be involved in effort of architecture conservation.

6 Conclusion

Heritage architecture carries the history and memory of a country. Appreciation and connection of present users to heritage architecture inculcate the value of nation within the people. Limited research has been done to discuss the architectural conservation of the heritage and its relationship to sustainability.

Due to the conditions during the time of construction of heritage buildings, regardless of vernacular architecture or colonial architecture, the materials and design consideration of such buildings are often still used as reference for the study of passive design strategy in many schools of architecture. Given such condition, it is clear that these buildings have already achieved the concept of sustainable architecture, what more if efforts are put into proper system of maintenance and conservation to ensure the structural integrity is retained. With that in mind, buildings of function of similar capacity could adopt and reuse heritage structures; thus, the need to clear green land for new construction in achieving sustainability in overall town planning is reduced.

Besides that, sustainability is more than just the environment, and it also comprises aspects of culture, society, and economy of a country. Architectural conservation on heritage building is an important symbolism to the people of our nation, as a reminder in the backdrop of our common identity in the multicultural context and entity of Malaysia. It is crucial to the effort of nation building, to strengthen the roots of the people with these physical representations of stories heard from the history books. On the other hand, heritage-based tourism is also one of the biggest economic drivers of Malaysia.

Green Building Index Council should also work with related heritage council in developing green building rating tool, specifically for heritage buildings. This guideline should take into consideration the aspects of cultural retention in the building to ensure the narrative and representation of our history and identity remain for the purpose of nation building so that sustainability is achieved in all aspects including culture, economy, and environment. This is because such guidelines and certifications will give better encouragement to the professions in the construction industry to take part in conservation of our heritage and use their respective expertise together. Further studies may occur to find a balance and solution in solving the issues faced in sustainable architectural conservation.

References

1. National Building Museum. (2012). *Historic Preservation Vs. Sustainability?* United States.
2. Harrington, J. T. (2004). *"Being here": Heritage, belonging and place making: a study of community and identity formation at Avebury (England), Magnetic Island (Australia) and Ayutthaya (Thailand)* (James Cook University). Retrieved from https://researchonline.jcu.edu.au/71/.
3. Thatcher, M. (2018). Introduction: The state and historic buildings: Preserving 'the national past'. *Nations and Nationalism, 24*(1), 22–42. https://doi.org/10.1111/nana.12372.
4. Uzzell, D. L. (1996). Creating place identity through heritage interpretation. *International Journal of Heritage Studies, 1*(4), 219–228. https://doi.org/10.1080/13527259608722151.
5. Harun, S. N. (December, 2011). Heritage building conservation in Malaysia: Experience and challenges. *Procedia Engineering, 20*, 41–53. https://doi.org/10.1016/j.proeng.2011.11.137.
6. KLRCA. (2014). Highlight: Wecome to Bangunan Sulaiman. *KLRCA Newsletter, 15*.
7. Badan Warisan Malaysia. (2011). *Conservation of Sulaiman*. Building, Kuala Lumpur.

Analysis of Road Traffic Services Due to the Operation of Cibubur Greater Jakarta Light Rail Transit Station

Rakhmat Shafly Syabana and R. Jachrizal Sumabrata

Abstract Cibubur is one of the areas developed by the railway transportation network system in the form of light rail transit. Light rail transit development is followed by the emergence of potential generation and attraction of the station's new trips. This study aims to analyze the influence of Cibubur LRT station operation on the services of the surrounding road network traffic and also proposes alternative development plans for the optimization of traffic services. In this study, the authors use VISSIM for the traffic models simulation. The double constraint gravity model is used to form the origin–destination (OD) matrix of the base year 2019. In determining the OD matrix of the LRT operational year 2021, the growth rate approach is used. The authors evaluate the simulation results of Cibubur LRT station according to PT. Adhi Karya master plan and the alternative scenarios are proposed. The results obtained showed that the operation of the Cibubur LRT station had a negative impact on surrounding traffic services. This can be seen from the average speed of the network which has decreased from 14 to 6–9 km/hr. Average network delay also increased from 191 to 351–733 s/vehicle. Average network stop also increases from 10 to 23–50 s/vehicle. And the stop average delay also increases from 75 to 144–393 s/vehicle. From the alternative development scenarios proposed, it was chosen to make flyover as direct access from and to Jl. Transyogi was selected as the best choice for optimizing road network services.

Keywords Traffic services analysis · Light rail transit · VISSIM · Double constraint gravity model · Cibubur LRT station

1 Introduction

Cibubur is a residential area for commuters working in Jakarta. Cibubur is a developing area that not only has housing but also has government activities, services, and business areas. In Regional Regulation No. 1 of 2015 concerning the Depok City

R. S. Syabana (✉) · R. J. Sumabrata
Faculty of Engineering, Department of Civil Engineering, Universitas Indonesia, Kampus Baru UI Depok, Depok 16424, Indonesia
e-mail: Rakhmat.shafly@gmail.com

S. S. Emamian et al. (eds.), *ICSDEMS 2019*,
https://doi.org/10.1007/978-981-15-3765-3_21

Spatial Plan for 2012–2032, Cibubur includes the area where the railway transportation network system will be developed. Cibubur is included in the area of intermodal service integration based on transit-oriented development (TOD). One of the efforts being made in developing the railway transportation network system is the construction of a light rail transit (LRT). The construction of LRT Jabodebek was initiated in 2015 through the Republic of Indonesia Presidential Regulation Number 98 of 2015 concerning the Acceleration of the Implementation of Integrated Light Rail in Jakarta, Bogor, Depok, and Bekasi. The construction of LRT in this area will certainly influence the surrounding area, one of which is the impact on traffic services. The impact on traffic services caused by the operation of LRT station in this area must be comprehensively studied. Therefore, this study was conducted to analyze the influence of the LRT station operational in the service of road network traffic around it. In addition, alternative development scenarios that can optimize traffic service conditions in the region are needed. In this study, the authors conducted a simulation to answer those needs.

2 Methods

This research was conducted to analyze road traffic services due to the operation of the Cibubur LRT station. The roads included in this study are Buperta Street, Taman Bunga Street, Jambore Street, Karya Bakti Street, Putri Tunggal Street, Jagorawi Toll Road, Alternatif Cibubur (Transyogi) Street, and Radar Auri Street (Fig. 1).

This study process can be seen in this flowchart below (Fig. 2).

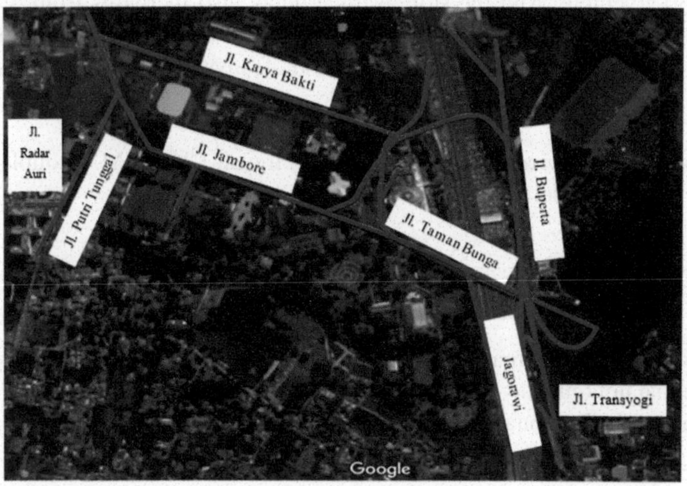

Fig. 1 Scope of studied road network (radius 800 m from station) [1]

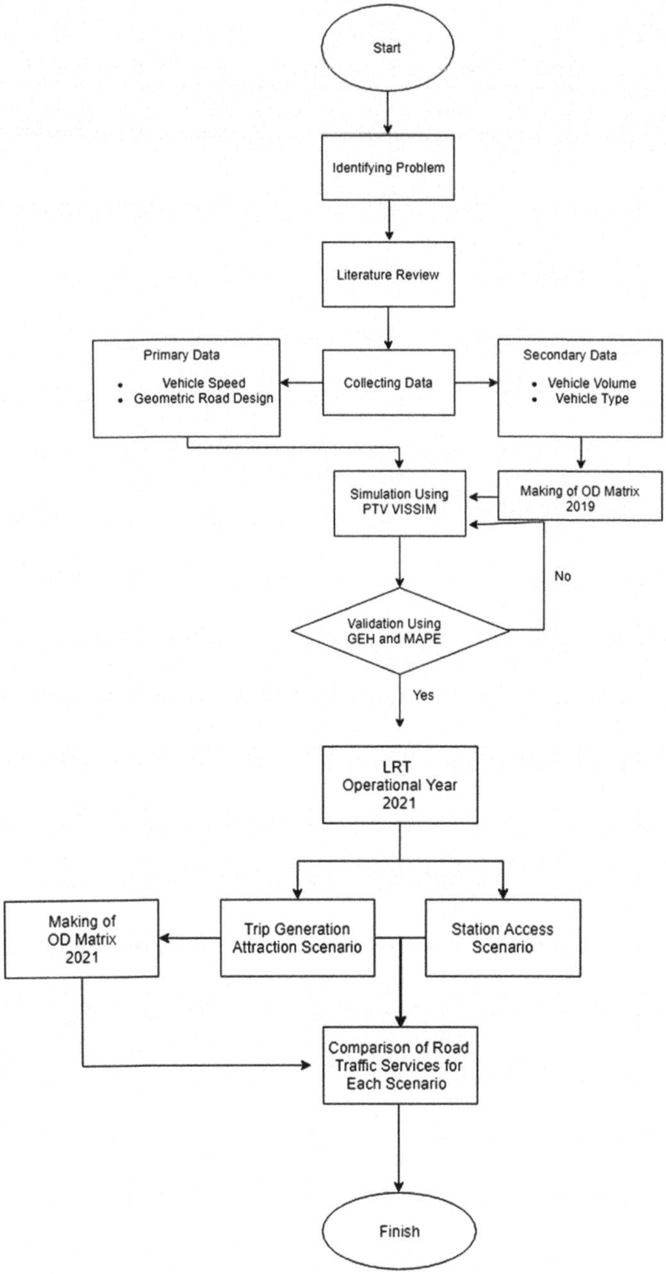

Fig. 2 Research process flowchart

Table 1 Zoning system

Zone	Base year 2019	Operational year 2021
Zone 1	Transyogi	Transyogi
Zone 2	Jagorawi (from Bogor)	Jagorawi (from Bogor)
Zone 3	Buperta	Buperta
Zone 4	Jagorawi (from Jakarta)	Jagorawi (from Jakarta)
Zone 5	Cibubur Junction	Cibubur Junction
Zone 6	Jambore	Jambore
Zone 7	Wiladatika	Wiladatika
Zone 8	Putri Tunggal	Putri Tunggal
Zone 9	Radar Auri	Radar Auri
Zone 10	–	Cibubur LRT station

3 Results and Discussion

3.1 Zoning System

Making the OD matrix required zoning system development in the existing year 2019 and the operational year of LRT station 2021. In this study, the origin–destination matrix was made with the constraint gravity model. The origin–destination matrix will be used as input in building traffic models at PTV VISSIM (Table 1).

3.2 Traffic Model Development of the Base Year 2019

In the development of the 2019 base year model, origin–destination matrix needed for each zone is defined in Table 2. The basic year OD matrix is obtained as follows.

The validation test was first performed with the GEH and MAPE statistical tests [2]. The results of the GEH and MAPE statistical tests obtained from this OD matrix are 0.083 and 0.23%. Therefore, the matrix is declared valid and can be used as input in modeling.

After the OD matrix is validated, modeling and simulation are carried out during rush hour on a Thursday afternoon. This is based on the peak busy time obtained from the results of the traffic counting survey conducted by IUTR. The simulation results in the form of speed on the road network around the LRT station are then also used for the validation of the model built, and validation is done by the same statistical tests, MAPE and GEH. The validation test conducted resulted in an average MAPE value of 8.22% and an average GEH of 0.379 so that the model built was valid and representative of the actual conditions. Therefore, road performance results from

Table 2 OD matrix of base year 2019

O/D	1	2	3	4	5	6	7	8	9	Oi (calculation)	Oi (survey)
1	0	115	25	1.017	116	1.319	44	411	100	3.149	3.156
2	762	0	16	438	101	568	13	177	43	2.117	2.122
3	82	22	0	4	5	87	0	13	3	217	217
4	2.578	677	213	0	21	82	1	73	18	3.664	3.672
5	73	19	4	85	0	156	4	49	12	401	402
6	1.488	4	11	854	197	0	52	695	170	3.471	3479
7	127	8	1	51	6	95	0	29	7	324	325
8	354	23	5	203	23	531	9	0	57	1.206	1.209
9	104	14	1	85	10	156	4	98	0	471	472
Dd (calculation)	5.568	882	276	2.738	479	2.993	127	1.546	411	15.020	
Dd (survey)	5.568	882	276	2.738	479	2.993	127	1.546	411		15.020

Table 3 Result of simulation for the base year 2019

No.	Parameter	Result
1	Speed average	14.529 km/h
2	Delay average	191.371 s/vehicle
3	Stop average	10.187 s/vehicle
4	Delay stop average	75.850 s/vehicle

model simulations can be accepted and the base year 2019 model can be used as a basis for modeling development in 2021 (Table 3).

3.3 *Traffic Model Development of the Operational Year 2021*

In the operational year of 2021, three different conditions were used related to the access of the LRT station. First, modeling is based on the access planned by PT. Adhi Karya and the remaining two conditions are alternative development scenarios offered to optimize road network traffic performance.

In modeling for the operational year 2021, demand variations are used because of the existing demand unknown. The use of demand variations as a plan for loading road traffic causes four OD matrixes to be formed for the operational year 2021. The formation of four OD matrixes is carried out with the Depok growth rate approach of 5.45% per year (compound interest) [3], and assuming the number of trips in the new zone of the LRT station is the result of shifting trips from the other nine zones (Table 4).

Table 4 Demand variation of road traffic loading

Headway (min)	LRT occupancy rate (%)	Number of passengers	Modal split (%)		Number of passengers by vehicle/h		Vehicle occupancy rate		Vehicle/h		
			Public transport (%)	Private transport (%)	Public transport	Private transport	Public transport	Private transport	Public transport	Private transport	Naming code
5	75	10.800	10	90	1.080	9.720	11	2	98	4.860	D-01
			90	10	9.720	1.080			884	540	D-02
10	50	3.600	10	90	360	3.240			33	1.620	D-03
			90	10	3.240	360			295	180	D-04

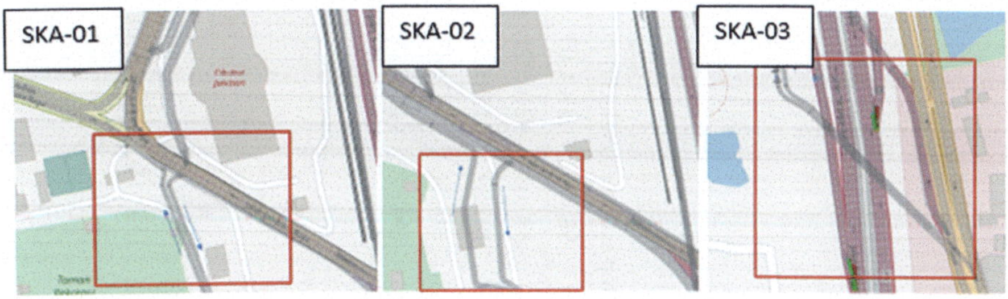

Fig. 3 Adhi Karya master plan (SKA-01), alternative scenario 1 (SKA-02), and alternative scenario 2 (SKA-03) [4]

The variation of demand is then used as input in making the model and was combined with each access condition as shown in Fig. 3. Then, the simulation is carried out to get the results of each measured parameter. The parameters used are speed average, delay average, stop average, and delay stop average (Table 5).

From the three models simulated, it can be concluded that in general, the performance of the SKA-03 model is the best of the three conditions. And from the simulation results, it can also be seen that Adhi Karya's access planning for LRT stations has a negative impact on the road network traffic services around it (Table 6).

Overall, the operation of the Cibubur LRT station makes the average speed in the road network decrease significantly.

4 Conclusions

Based on the results of the research and analysis that have been carried out, we can conclude a number of things as follows:

1. In the basic year 2019, the condition of the road network in the Cibubur area during the afternoon rush hour has an average speed of 14.529 km/h, a delay of 191.371 s/vehicle, stop average of 10.187 s/vehicle, and a delay average of 75.850 s/vehicle.
2. Operations of the Cibubur LRT station with a station access plan based on the planning of PT. Adhi Karya caused the average road network speed at the afternoon rush hour to be 5–9 km/h, delay average 351–733 s/vehicle, stop average 23–50 s/vehicle, and delay stop average 144–393 s/vehicle.
3. The best alternative scenario for optimizing the performance of the road network is to make LRT station access directly from and to Transyogi Street (SKA-03).
4. The speed of the vehicle in the station access road is affected by modal split. More and more public vehicles used in the station make the speed on the access road increase.

Table 5 Results of model simulation

Demand	Delay average (s/vehicle)			Stop average (s/vehicle)			Speed average (km/h)			Delay stop average (s/vehicle)		
	SKA-01	SKA-02	SKA-03	SKA-01	SKA-02	SKA-03	SKA-01	SKA-02	SKA-03	SKA-01	SKA-02	SKA-03
D-01	733.14	611.63	412.43	50.95	39.61	26.59	5.04	6.06	7.84	393.04	330.08	210.221
D-02	418.56	422.67	356.22	28.68	27.99	22.16	8.41	8.40	9.27	179.01	187.29	158.655
D-03	433.23	451.68	349.57	31.58	32.15	20.44	8.21	7.99	9.43	175.44	191.38	162.847
D-04	351.97	342.94	263.32	23.91	22.67	12.79	9.55	9.71	11.69	144.46	144.87	124.311

Table 6 Comparison of performance of models with existing models

Scenario	Demand	Delay average (%)	Stop average (%)	Speed average (%)	Delay stop average (%)
SKA-01	D-01	283.100	400.227	−65.295	165.751
	D-02	118.721	181.580	−42.056	53.909
	D-03	126.388	210.062	−43.439	52.042
	D-04	83.923	134.741	−34.250	35.854
SKA-02	D-01	219.607	288.901	−58.265	132.851
	D-02	120.866	174.756	−42.137	58.236
	D-03	136.027	215.597	−44.957	60.374
	D-04	79.204	122.564	−33.143	36.070
SKA-03	D-01	115.514	161.025	−46.040	70.217
	D-02	86.144	117.551	−36.189	43.270
	D-03	82.668	100.706	−35.037	45.459
	D-04	37.599	25.566	−19.536	25.325

Acknowledgements This article is a part of research of International Indexed Publication for Universitas Indonesia Student's Final Project (PITTA) 2019.

References

1. Calthorpe, P. (1993). *The Next American Metropolis.* Los Angeles: Princeton Architectural Press.
2. Hellinga, B., Rakha, H., & Aerde, M. (1996). *Systematic Verification, Validation, and Calibration of Traffic Simulation Models.* Washington DC: Transportation Research Board Annual Meeting.
3. Dinas Perhubungan Kota Depok. (2017). *Laporan Akhir Kajian Kinerja Jaringan Jalan di Kota Depok 2017.* Depok.
4. PTV Group. (2018). *PTV VISSIM User Manual.* Karlsruhe: PTV Group.

The Analysis of Pedestrian Scenarios in Public Spaces (Case: Harjamukti LRT Station, Cibubur)

Lucky Steffano Samuel Situmorang and R. Jachrizal Sumabrata

Abstract The development of an area must meet various supporting aspects so it can develop rapidly, one of them being transportation. Transportation as the lifeblood of a country's political, economic, social, cultural, defense, and security life should be considered by the government so the purpose of transportation use will be successful. In 2015, through Presidential Regulation (Perpres) Number 98, the government-appointed PT. Adhi Karya Tbk. to carry out the construction of Jabodetabek LRT (Light Rapid Transit) infrastructure with one of them being the Cibubur-Cawang route Cibubur Station must be noticed with the rights of its users. LRT passengers should be moving within the scope of the station building on foot. This research is intended to analyze the level of service of pedestrian scenarios at LRT stations when they are completed and being operated. The research was conducted by simulating pedestrian models at the Station with plan design using Viswalk PTV software. Model parameters are estimated based on the parameters of the target passenger plan conditions. This research resulted that station design in the plan and alternative scenarios is still able to accommodate the number of passengers planned with a few notes. The level of pedestrian service for areas, ramps, and queues in alternative scenarios tend to be better than the planned scenario. Anticipation needed when the station has been operating due to various possible scenarios that affect the level of service of LRT Harjamukti Station, Cibubur.

Keywords Level of service (LOS) · Simulation · PTV viswalk · LRT station · Area · Ramp · Queuing · Scenario · Alternative

1 Introduction

Transporation as a life nerve of politic, economic, social, culture, safety, and defence should be the first attention by the government so that the function of it will be successful. Transportation network system success is assessed by its effectivity. This

---

L. S. S. Situmorang (✉) · R. Jachrizal Sumabrata
Department of Civil Engineering, Faculty of Engineering, Universitas Indonesia, Kampus Baru UI Depok, Jakarta 16424, Indonesia
e-mail: luckysitum@gmail.com

S. S. Emamian et al. (eds.), *ICSDEMS 2019*,
https://doi.org/10.1007/978-981-15-3765-3_22

system should guarantee user safety, accessibility, integration, capacity, orderliness, punctuality, and other aspects. As an Indonesian government effort in developing mass transportation infrastructure in the capital city, they assign PT Adhi Karya Tbk through Peraturan Presiden (Presidential Regulation) No 98 the year of 1995 to build LRT infrastructure. LRT development plan is expected to provide opportunities of development of the surrounding area. One of the routes to be built is Cawang-Cibubur route aiming to reduce traffic congestion, which has become a problem by expecting people to use public transportation. This route is expected to be completed at the end of 2019. This route will have the Harjamukti Station as the end platform.

One of the transportation components within the Cibubur LRT station is pedestrians who have used and will use the train. The rights are regulated and protected in UU Number 22 Year 2009. In this regulation, giving priority for safety and comfort in traffic management and engineering has been stated. The protection and fulfillment of pedestrian rights must be considered when planning and building public facilities. In this research, the study case is Cibubur LRT station where users in the building will walk on foot. This research aims to analyze the level of service of pedestrian facilities in public spaces at the Cibubur LRT station when it is operating.

2 Literature Review

A 'pedestrian' is a person on foot, or in or on a contrivance equipped with wheels or revolving runners that is not a vehicle. This can include an able pedestrian, a person pushing a pram, a person on a skateboard, a person in a wheelchair, and a number of other users [1]. Pedestrians are also people who are in the early and final phases of motorized travel; the driver or passenger walk towards the vehicle when they are parking, drive to their destination, park their vehicle, and walk to their final destination. In urban centers, pedestrian flows can be very significant and must be accommodated in the planning and design of traffic, facilities, and signs. Pedestrian safety is also a big issue because pedestrians are at a disadvantage when there is a potential for vehicle conflict with pedestrians, for example at intersections [2].

The level of service (LOS) of pedestrian facilities is intended to determine the qualitative aspects of the design that has been made. Designers of pedestrian facilities must pay attention to all elements such as pedestrian characteristics and platoon formation due to traffic light cycles or queues [3]. The service level standard used is based on the number of people at large. The level of pedestrian service on a pedestrian path is divided into six levels based on the range of modulus of pedestrian space filled (Table 1, [6]).

Table 1 Pedestrian level of service

Level of service	Pedestrian space (m^2/p)		
	On pedestrian path	On stairs	When queueing
A	>5.6	>1.9	>1.2
B	3.7–5.6	1.6–1.9	0.9–1.2
C	2.2–3.7	1.1–1.6	0.6–0.9
D	1.4–2.2	0.7–1.1	0.3–0.6
E	0.75–1.4	0.5–0.7	0.2–0.3
F	<0.75	<0.5	<0.2

3 Methods

The methodology of analyzing the scenario of pedestrian services at the Cibubur LRT station begins with the formulation of the research problem, then the researcher conducts a literature review to determine the variables needed for the simulation. Next, the researcher collected secondary data from the Harjamukti LRT station developer, which is the target number of services and the geometry of the layout of the planned location. After the data collected, the researchers made a simulation model using PTV Viswalk software and ran the simulation. In developing the model, it is necessary to define the inputs, processes, and desired outputs. Simulation models may utilize physical designs, such as networks or facilities, and often require collection of data, sometimes in considerable details, and their translation in a form that can be put into mathematical formulae and relationships of the program [4]. Pedestrians can only move on the designated areas and ramps, hence, those areas are created on all walkable locations. If there is a location in the area where pedestrians do not or cannot move, an obstacle can be created on top of the pedestrian area. Construction of several smaller areas rather than one large area is recommended in order to minimize the computational efforts [5]. The simulation that is run then needs to be validated to prove that the model created can describe real conditions. After the simulation is stated it can describe the real conditions, then researchers analyze the simulation results according to the level of service. From the results of the analysis, conclusions can be drawn.

Determination of the variables used in this study was determined based on a literature review conducted by researchers. These variables are the pedestrian volume and the LRT travel headway.

4 Data Analysis and Discussion

4.1 Forecasting Pedestrian Movements

The built model requires input data which are variables that change to affect the model scenario. The variables are developed into:

1. Headway variable for LRT trips

In this model, two scenarios are determined based on the headway variable, which is 5 min as the planned headway and 10 min as an alternative headway.

2. Number of alighting pedestrians variable

The number of alighting pedestrians depends on the capacity of the train line that stops at the station. The normal capacity of the LRT with 6 cars is 708 passengers and the maximum capacity is 1248 passengers.

3. Number of boarding pedestrians variable

Passengers who board or want to depart on the model will start from the plaza area on the ground floor. The appearance of pedestrians in the model is determined on the Input Pedestrian, which is placed in the plaza area. The volume of pedestrians departing is referred to by the capacity of the train and headway circuits. The number of passengers who will travel (boarding) is obtained from the following formula.

$$V = Oc \times \frac{60}{H} \times n$$

V Total number of passengers (passenger/h)
Oc Normal capacity of train circuit (passenger)
H Headway train series (min)
N Occupancy Factor (%).

From the three variables mentioned above, each scenario is combined to produce four planned scenarios and four alternative scenario simulations. The following scenarios are arranged as follows (Table 2).

4.2 Simulation Results

Pedestrian level of service for areas in the simulation is using parameters based on AASHTO, 2000 for numerical data processing and color diagrams in the simulation.

Figures 1, 2, and, 3 show the color diagram of the service level of each unpaid area, north and south paid areas, and west platform for each of the plan scenarios and alternative scenarios. For all the areas mentioned, there has been a significant

Table 2 Combination of modeling scenarios

Scenarios		Boarding		Alighting		Headway (min)
		LRT occupancy factor (%)	Total passengers (passanger/h)	LRT occupancy factor (%)	Total passengers (passangertrain)	
Plan	1	50	7.488	50	624	5
	2	75	11.232			
	3	50	7.488	75	936	
	4	75	11.232			
Alternative	5	50	3.744	50	624	10
	6	75	5.616			
	7	50	3.744	75	936	
	8	75	5.616			

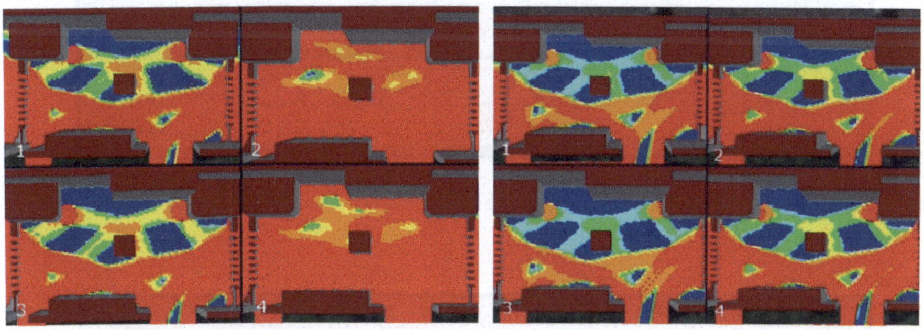

Fig. 1 Color diagram of unpaid area plan scenario (left) and alternative scenario (right) 1–4 under maximum density conditions

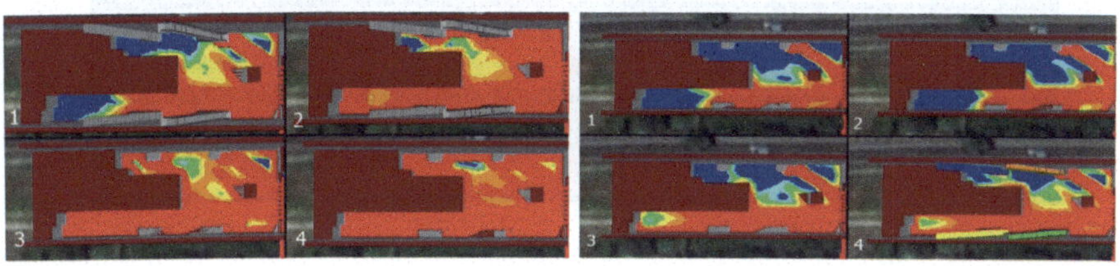

Fig. 2 Color diagram of North paid area plan scenario (left) and alternative scenario (right) 1–4 under maximum density conditions

increase in the level of service areas with the expansion of service area A and the reduction in service area F in the area concerned. All service level color diagrams were in maximum density conditions when the simulation was run (Figs. 4 and 5).

In the level of service color diagram of the east platform for the plan and alternative scenarios, almost all of the areas show the level of service F. Alternative scenarios

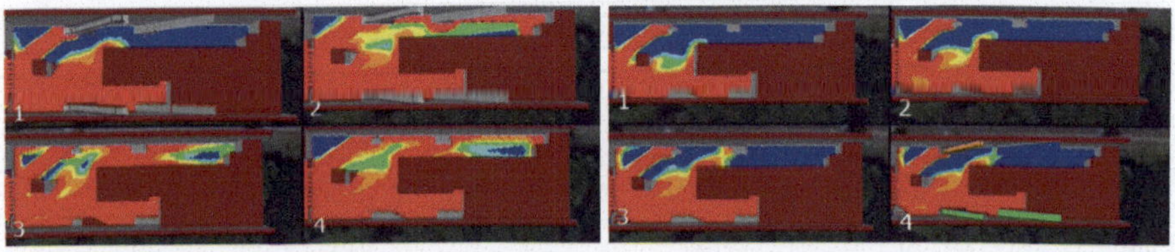

Fig. 3 Color diagram of South paid area plan scenario (left) and alternative scenario (right) 1–4 under maximum density conditions

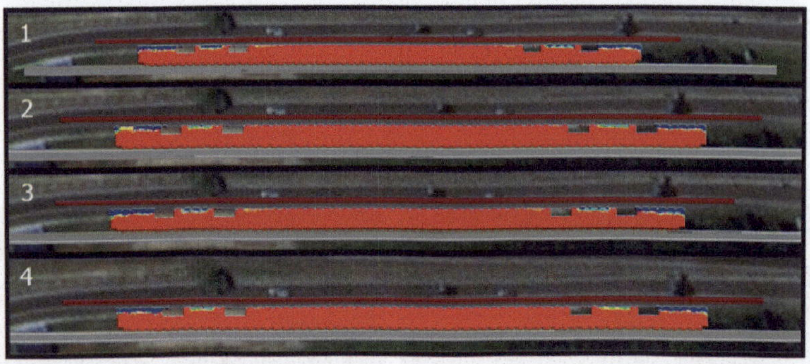

Fig. 4 Color diagram of the Eastern platform area plan scenario 1–4 under maximum density conditions

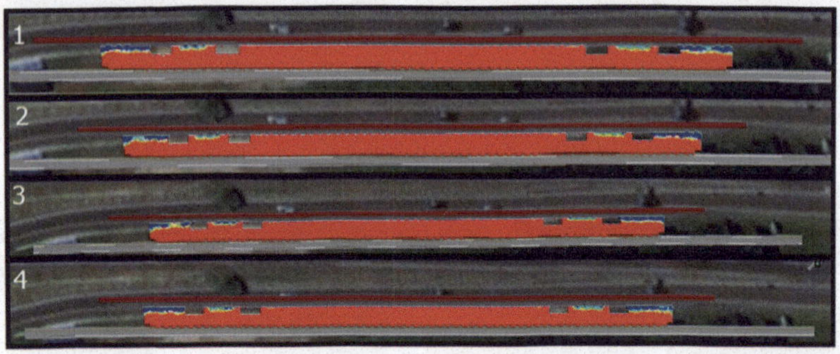

Fig. 5 Color diagram of the Eastern platform area alternative scenarios 1–4 under maximum density conditions

even increase the density of pedestrians waiting on the east platform, which is due to the longer train headway which makes pedestrians increasingly pile up on the east platform waiting for the train to be arrived.

The level of service on the stairs and escalators in all building components shows a significant increase between planned and alternative scenarios. Alternative scenarios can increase the average modulus of pedestrian space on stairs and escalators.

Table 3 Combination of modeling scenarios

Scenarios		Area	DensAvg (Ped/m^2)	Average Space Modulus	Level of Service	DensMax (Ped/m^2)	Max Space Modulus	Level of Service
Plan	1	South Tap Gate Out	0,601	1,66343	LOS D	1,747	0,572426	LOS F
	2	South Tap Gate Out	0,525	1,905538	LOS D	1,817	0,550315	LOS F
	3	South Tap Gate Out	5,265	0,189919	LOS F	13,773	0,072607	LOS F
	4	South Tap Gate Out	5,339	0,187287	LOS F	13,055	0,076597	LOS F
Alterna-tive	1	South Tap Gate Out	0,263	3,801775	LOS B	1,722	0,580627	LOS F
	2	South Tap Gate Out	0,281	3,560588	LOS C	1,691	0,59138	LOS F
	3	South Tap Gate Out	0,661	1,513339	LOS D	2,513	0,397938	LOS F
	4	South Tap Gate Out	0,646	1,54704	LOS D	2,529	0,395474	LOS F

However, for the comparison between stairs and escalators, stairs tend to increase in greater modulus of pedestrian space compared to escalators. This is because pedestrians tend to prefer ramps that require less effort to move so that when the level of escalator service is slightly better, pedestrians who initially want to choose stairs will switch to using escalators because of less travel time and effort.

The level of service in the tap gate queue for the entire south and north sides and for in or out routes also increased in alternative scenarios compared to the planned scenario. Except for the southern tap gate out, the level of queue service at maximum density conditions does not increase at all and the modulus increase in pedestrian space is not significant. The service level of the southern tap gate out is shown in the table (Table 3).

This is because the queue space in the southern paid area is smaller than the queue space in the northern paid area. In the plan layout, the southern paid area is smaller because there are more rooms that are used for station offices and others. Its level of service cannot be increased and it must be considered carefully, given the level of service in all scenarios when the maximum density condition is the worst.

5 Conclusions

Based on the results of the research and analysis that have been carried out, we can conclude a number of things as follows:

1. Alternative scenarios when the headway of train is 10 min are better solutions than the ones that have been run in the simulation. It shows an increasing level of service in several aspects, namely area, ramp, and queue, compared to the plan scenario, where F service levels are more common in various areas.
2. In areas such as first floor plazas, access bridges, north and south pay areas, non-paid areas, and west platforms service levels in alternative scenarios have been increased. However, on the east platform, for an alternative scenario, the service level of the eastern platform area remains at LOS F, which is due to the

increasing train headway, making pedestrians pile up on the platform waiting for the train series to be available on the platform.

3. For the level of pedestrian services on stairs and escalators on the first and second floors, they tend to increase in alternative scenarios compared to the planned scenarios.
4. In the southern tap gate out queue, the level of pedestrian services tends to be worse than the northern tap gate out even though it has been implemented in both scenarios. This tendency occurs due to the queuing space on the south side which is narrower due to the use of space as station office space which is more than the north side.

Acknowledgements This article is a part of the research of International Indexed Publication for Universitas Indonesia Student's Final Project (PITTA) 2019.

References

1. NZ Transport Agency. (2008). Pedestrian planning and design. Wellington.
2. Mathew, D. T. (2014). *Transportation Systems Engineering*. Mumbai: IIT Bombay.
3. Fruin, J. J. (1971). Designing for Pedestrian: A Level-of-Service Concept.
4. Galiza, R., Ferreira, L., Kim, I., & Laufer, J. (2011). Modelling pedestrian circulation in rail transit stations using micro-simulation.
5. Beutin, T. (2012). PTV Viswalk realistic simulation of pedestrian flows.
6. American Association of State Highway and Transportation Officials. (2000). Highway Capacity Manual. Washington: National Academies of Science.

The Changes of Facades in Jengki Building with Commercial Function in Yogyakarta

Noor Zakiy Mubarrok and Sidhi Pramudito

Abstract Jengki is one of the architectural styles that was developed and spread in several regions of Indonesia between the 1950s and 1960s. Buildings with Jengki architectural style have their own values and identity as part of the development of authentic modern architecture in Indonesia. Until now, the existence of buildings with Jengki architectural style still can be found in Indonesia, one of them is Yogyakarta. But over time, the existence of buildings with Jengki style has changed because of several interests. The Jengki buildings often to be abandoned, they become unmaintained, damaged, or sold and then turned from residential into another function. This condition also occurs in Jengki buildings in Yogyakarta. In this research, the authors try to study the façade changes of Jengki buildings due to the function changes as commercial buildings. Through the research, it is expected to be able to identify which of the element on the facades of Jengki architectural style is still surviving, change, or disappear, and also how far is it still recognized as a Jengki building. This is qualitative-descriptive research with a purposive sampling technique. The building samples categorized in Jengki architectural style, then analyzed based on phenomena, especially in façades element which currently function as commercial buildings. Based on the analysis, it was found that the changes in the façade elements mostly occur in buildings that have functional change from the residential to commercial. The addition or changes in the façade element make the characteristic of the Jengki architecture less visible.

Keywords Changes · Façade · Jengki · Commercial

N. Z. Mubarrok (✉) · S. Pramudito
Department of Architecture, Universitas Atma Jaya Yogyakarta, Yogyakarta, Indonesia
e-mail: noor.zakiy@uajy.ac.id

S. Pramudito
e-mail: sidhi.pramudito@uajy.ac.id

S. S. Emamian et al. (eds.), *ICSDEMS 2019*,
https://doi.org/10.1007/978-981-15-3765-3_23

1 Introduction

Jengki is an architectural style that was developed and spread in several regions of Indonesia in the 1950s–1960s [1–3] in [4]. According to some literature, there is a mention that Jengki is a modification and not an advanced stage of the previous architectural style, The Dutch Colonial Architecture [5]. Other sources also mentioned, although synchronically not contemporaneous, the spirit of Jengki that emerged could be called a typical Indonesian style [1]. Jengki is also said to be the style of Indonesian's modern architecture after independence.

As an architectural style, buildings with Jengki style are basically more dominated with functional purposes related to the tropical climate. Such as the slope of the roof is rather steep to facilitate the flow of rainwater, the formation of a pentagonal that extends upward on the wall for sun shading, the terrace to reduce the heat of the indoor area, and the roster to providing natural air inside buildings [3]. But Silas [1] said that the Jengki is also an architectural expression that illustrates the spirit of freedom and independence from colonial politics in Indonesia around 1950s–1960s. Therefore, in fact, buildings with Jengki architectural style have their own values and identities as part of the development of modern authentic architecture in Indonesia. The expression of freedom then combined with the functional aspect related to the tropical climate, which becomes the main priorities in the Jengki buildings to create a distinctive design.

Jengki as a modern Indonesian architectural style certainly has distinctive characteristics, especially in its facade. The "eccentric" form mark the spirit of freedom from colonialism in the past. Some of the characteristics of the facade elements of Jengki buildings [3] in [6] are: The type of roof that is commonly used in Jengki buildings is a gabble roof with a slope of approximately 35°, with different height, Concrete folded structure above the terrace, Pentagon wall at the front of buildings, roster as an aesthetic element not only for providing natural air inside the building, and asymmetrical composition of doors and windows in the façade.

Nowadays, the existence of the buildings with Jengki style can still be found in Indonesia, for example in Malang, Surabaya, Lawang, and Bandung (documented through several scientific articles). But over time, the existence of buildings with Jengki style has changed because of the demands of time. Its existence can even be said to be threatened with disappearing because more and more buildings are not used, old, not maintained, sold by the owner then the new owner changed the functions and its form. And in the end, the buildings with Jengki style will become a rare piece of architecture.

The same condition also occurs in Jengki buildings in Yogyakarta. Some buildings were abandoned by their owners and are poorly maintained. However, some of these buildings have changed its functions into other forms of private buildings (residential houses) into limited public buildings (commercial). The good side is that with changes in function, indirectly, the Jengki buildings will continue to be preserved and its existence will remain in the present. In this research, the authors try to study the changes in building facades with Jengki style in Yogyakarta. Through the façade some information related to the culture since the buildings were built can be gathered. Facade can also uncover the criteria of order and arrangement and is credited with providing possibilities and creativity in ornamentation and decoration [7].

Facade becomes important in architecture because from façade, an architectural work can be recognized for its function, style, and even its meaning. The facade is an element that connects the exterior and interior in architectural works, an element in the enclosure of a building that has a meaning as the face of architectural work [8] in [9]. Facades cannot be separated from architectural works. The facade is an element that records the history of human civilization [10] in [9].

Some of the facade elements are roofs, walls, windows, doors, and floors [11], while Krier [7] describes in more detail about the elements forming the facade, which consists of: entrance, the entrance is a transition area between exterior and interior. An element of self-declaration of building occupants. The second is floors: the floor is a building base, an element that has strong ties with urban areas. This is because the floor has a direct connection to the ground/road. Commercial buildings and offices often accommodate floors as part of identity. The third is windows and doors: the window is related to the breadth of the view for the occupants in the building, ensuring the penetration of sunlight into the room, as well as being a decorative element in the facade. Doors have an important role in buildings. An element that allows humans to penetrate a vertical wall, which can be made firm like a gate, or can be the simplest way by punching holes in the wall. Placement of the entrance is also related to the circulation of patterns of activity in space. The fourth is a fence (barrier): is a physical boundary that separates the use of space. The fifth is roofs: This is the element that is at the top of a building. Also, regarding the skyline of buildings, related to the context of the surrounding environment. And the last is sign and ornament: as a sign installed by the building owner, providing building identity information to the public. Ornaments are a complement to the visual aesthetics of the building. As a decorative element, the attraction to attract public attention.

Over time, the facade of an architectural work can change. This change can result in the erosion of architectural identity, and can even erode the identity of an area/region. Economy factors, social, and cultural factors, as well as environmental ecology, are the main factors that influence the changing of a facade [9].

The purpose of this research was to determine changes in the facade of the Jengki building where the building currently functions as a commercial building. It is also hoped that specific architectural elements of Jengki style can be identified which still survives or disappear.

2 Methods

This research is included in a qualitative-descriptive study with a purposive sampling technique. Building samples categorized in the Jengki architectural style, then analyzed based on phenomena, through the façade elements: entrance, floors, windows, doors, fences, roof, sign and ornaments [7] to see how far the façade changes in the commercial building with Jengki architectural style, and also how far is it still recognized as a Jengki building.

3 Result and Discussion

The research sample located in the commercial area in Yogyakarta. First sample in the Jl. Cik Di Tiro as a Café, the second sample in the Jl. Kartini as a Course place and the third located in the Jl. Brigjend Katamso as a franchise restaurant. These three sample objects can be accessed easily because they are located near the downtown of Yogyakarta.

3.1 1st Sample Object; Coklat Café

Located in Jl. Cik Di Tiro no. 19, Gondokusuman, Yogyakarta, been a café since 1994. Several architectural elements can be identified as the characteristic of Jengki; the folded concrete structure above the terrace, natural tone materials as decorative façade, various applications of aesthetic elements (trellis pattern, fence), and various window shapes with asymmetrical composition.

The existence of the café function in the Jengki building since 1994 has resulted in several changes to the building façade. Analysis of changes that occur in the building façade elements is as follows:

3.1.1 Entrance

The entrance to the site is designed without borders to show openness as a commercial building. Entrance to the building was changed by adding enclosure on the terrace as an expansion area, with doors and glass windows in aluminum frame.

3.1.2 Floor

The site is located higher than the road, parallel to the pedestrian area, with a concrete block, as a visitor parking area, and an outdoor café area at night. The floor of the building is higher than the parking area as seen by the stairs to enter the building.

Fig. 1 Changes to the Coklat cafe entrance building

3.1.3 Windows and Doors

The windows and doors at the entrance of the building were changed due to the expansion area on the terrace. The old window is still maintained, but the old entrance is demolished to expand the indoor area (Fig. 1).

3.1.4 Fence

The café owner changes the fence by dismantling the original fence. The café design has no barriers with the pedestrian ways, and makes it open to attract visitors.

3.1.5 Roofs

The Coklat café retains the original roof, the limasan. Changes occur with the addition of a canopy with zinc and polycarbonate material at the front of the building, as a sun shading and expansion of the outdoor cafe area. The installation of the canopy adjusts to the existing concrete folded structure in the facade.

3.1.6 Sign and Ornaments

The building sign is placed separately from the building, which is in the building yard, in the form of a vertical single sign (billboard) so it does not cover the building appearance. The original ornament that is still preserved is Javanese letters (Wisma Badra Naya) using metal material that describes the building owner, located on the front wall of the building, which is emphasized against the background of black natural stone (Fig. 2).

Fig. 2 Changes to the fence, canopy additions and ornaments in the Coklat cafe building

3.2 *Sample Object; Neutron Building*

Neutron Building, a tutoring institution located at Jl. Kartini No. 1 Gondokusuman Yogyakarta. It has the characteristics of Jengki style which can be seen from the shape of the gable roof with a slope of approximately 35°, concrete folded structure above the terrace. Asymmetry composition in the facade can be seen from the articulation of the facade, which is strengthened by the use of different natural stone materials as decorative elements on the facade.

This building often changes its function, among others, into offices and cafes. Since 2018 this building has been used by Neutron as a tutoring institution. Analysis of changes that occur in the building facade elements:

3.2.1 Entrance

The entrance to the site has not changed while the entrance to the building is converted into a 12 mm tempered glass door, frameless, give a welcoming impression to the visitors.

3.2.2 Floors

The site is at a higher level compared to the road, with concrete block for the visitor parking area. The terrace and the indoor area still maintains the original level and material.

3.2.3 Windows and Doors

The main windows and doors of the building have been changed by Neutron, using tempered glass, 12 mm frameless with the position still maintaining the composition of the original building façade to give a welcoming impression to the visitors.

Fig. 3 The main entrance and parking area of neutron building

3.2.4 Fence

Neutron is still maintaining the original fences of the building (Fig. 3).

3.2.5 Roofs

The main roof of the building and also the concrete folded structure above the terrace does not change.

3.2.6 Sign and Ornaments

Billboard located in the front of the site, while banners hanging in the concrete folded structure. The banner position makes the concrete folded structure less visible (Fig. 4).

3.3 2nd Sample Object; Olive Chicken

Olive Chicken, a franchise restaurant located on Jl. Brigjen Katamso, Yogyakarta. This building was originally a house and shop building, seen from the proportion of openings in the facade. The left is a shop and the right is the entrance to the house.

Fig. 4 The sign and ornaments at neutron building

Fig. 5 The entrance and floors in olive chicken

Been used by Olive Chicken since 2018. It has the architectural characteristics of Jengki seen from the use of a saddle roof with a slope of approximately 35°, with the position of the roof ridge that jutted forward, the concrete folded structure above the terrace, asymmetry composition of the openings in the façade, blinds, and roster to maximize natural ventilation above the window frame. Description of changes in this building is as follows:

3.3.1 Entrance

The entrance does not change since the building was built. The sliding doors for the shop and entrance to the house are still maintaining the original material. The divider between the shop and house area was removed to accommodate more visitors.

3.3.2 Floors

This building does not have a terrace nor a yard. It is directly adjacent to pedestrian ways. The level of the building's floor is slightly higher than pedestrian ways with white ceramics tiles (Fig. 5).

3.3.3 Windows and Doors

The windows and doors do not change since the building was built. Wood sliding door for the shop and wood door and window to entrance the house.

3.3.4 Fence

This building does not have a fence, it is directly adjacent to the pedestrian ways.

Fig. 6 The entrance and floors in olive chicken

3.3.5 Roofs

The main roof of the building does not change, as well as the concrete folded structure above the terrace. There is additional galvalume roof below the concrete folded structure to reduce the penetration of sunlight and rainwater. This addition makes the compositions of the roster in the façade not visible.

3.3.6 Sign and Ornaments

A signboard placed at the top of gable wall, which is blocking the blinds at the gable wall. Banner hanging at the additional roof of the façade (Fig. 6).

Overall, based on the analysis of each sample object, some propensities in structuring the façade are as follows (Table 1).

4 Conclusions

Based on the analysis of the three sample objects, the most changes of the building façade most occurred in samples 1 (Coklat Café) and 2 (neutron), especially in the type and material of openings (doors and windows). This is due to the fact that the two buildings changed their function from a residential function (since the building was built) to a commercial function. In contrast to sample 3 (Olive Chicken) does not change its function since the building was built, but the divider of the indoor area was removed, to accommodate more visitors. The addition elements such as banner/sign on the façade, and additional canopy make the characteristic of Jengki architecture less visible. It is considered for the owner/tenant of the building about the façade composition when adding a new element to it, so the new addition element does not interfere with the visibility of the Jengki architectural characteristic.

Table 1 Comparative analysis of facade change in the research samples

Façade elements [7]	Jengki building sample category		
	Coklat café	Neutron	Olive chicken
Entrance	Changed in part due to area expansion	Changed material at the entrance of the building	Does not changed
Floors	Changed in the function of outdoor space into parking area with concrete and grass divider	Changed in the function of outdoor space into parking area with concrete block	Changed in floor material in indoor area with 40 × 40 ceramics tile
Windows and doors	Changed in part: There are additional doors and windows on the terrace due to the expansion of the area	Material changed using frameless tempered glass door and window	Does not change
Fence	Changed: several fences are removed for openness impression	Does not change	Without fence
Roofs	The main roof does not change. Canopy was added because of the expansion area, covering the concrete folded structure	Does not changed	Main roof and concrete folded structure do not change. Galvalume canopy was added above the openings
Sign and ornaments	Billboard placed separately from the building, it does not cover the building façade	Billboard placed separately from the building, there are additional banners attached to the eaves	Sign attached in the top gable wall, additional banner attached in the galvalume canopy

References

1. Prijotomo, J. (1996). When west meet east: One century of architecture in Indonesia (1890s–1990s). *Architronic*, 2–10.
2. Sukada, B. (2004). Langgam jengki langgam khas Indonesia (disarikan dari bahan kuliah Ir. Budi Sukada Jurusan Arsitektur UI). Tabloid Rumah 20 Juli–02 Agustus 2004, Jakarta.
3. Widayat, R. (2006). Spirit dari rumah gaya jengki ulasan tentang bentuk, estetika dan makna. *Dimensi Interior, 4*(2), 80–89.
4. Wulandari, R. (2016). Bandung Jengki from heritage point of view documentation and preliminary search on significance. *Dimensi, 43*(1), 55–66.
5. Setyabudi, I. (2012). Tipologi dan morfologi arsitektur rumah jengki di kota malang dan lawang. *Arsitektur e-journal universitas brawijaya program magister dan doktoral program studi arsitektur lingkungan binaan, 5*(1), 32–46.
6. Mubarrok, N. Z., Pramudito, S. (2018). A study of façade element on jengki architecture in Yogyakarta in the 1st ICCS and the 9th ISVS reframing vernacular: Politics, semiotics and representation. Denpasar.
7. Krier, R. (1988). *Architectural composition*. New York: Rizolli.
8. Moloney, J. (2011). Designing kinetics for architectural façade. State Change.

9. Setiawan, D., & Utami, T. B. (2016). Tipologi perubahan elemen fasad bangunan ruko pada penggal jalan puri indah Jakarta. *Vitruvian, 6,* 15–24.
10. Kamurahan, R. S., Waani, J. O., & Rogi, O. H. (2014). Studi persepsi masyarakat terhadap estetika desain fasad bangunan dengan pendekatan teori subyektif studi kasus di koridor boulevard on business (bob) jalan piere tendean manado. *Media Matrasain, 11,* 68–82.
11. Arif, B. (2014). Perubahan fasade rumah di perumahan woningpark gergaji semarang. *Teknik, 35*(1), 1–7.

Bearing Strength of Reinforced Concrete Blocks Axially Loaded Through Various Sizes of Steel Plates

Mohd Raizamzamani Md Zain, Norrul Azmi Yahya, Siong Wee Lee, Chai Lian Oh, and Anizahyati Alisibramulisi

Abstract This paper presents an experimental investigation into the bearing strength of concrete blocks axially loaded through various sizes of steel plates. The dimension of the loaded surface of concrete block specimens relative to the bearing steel plate is one of the most significant factors influencing the bearing strength of concrete blocks. Hence, the main purpose of this study is to determine the ultimate bearing strength and observe the modes of failure of concrete blocks axially loaded through various sizes of steel plates. A total of twelve (12) numbers of specimens with three different sizes of steel plates on top of concrete blocks were experimentally investigated. The results of the experimental testing herein were compared with the values of bearing strength calculated based on the previous studies. Experimental results demonstrate that the different sizes of steel plates depicted a different kind of crack failure for the concrete blocks. The enveloping concrete provides an effect on the increment of concrete compressive strength at bearing area.

Keywords Concrete blocks · Steel plates · Bearing strength · Axially compressive loading

1 Introduction

In general, concrete is a mixture of coarse (made of crushed stone) and fine aggregates, cement (such as Ordinary Portland Cement), and water. With the use of reinforced concrete, compressive strength can be improved. As presented by numerous numbers of studies [1–3], several crack patterns such as vertical and radial cracks were observed on the unreinforced concrete blocks due to high compressive loading.

M. R. Md Zain (✉) · N. A. Yahya · C. L. Oh · A. Alisibramulisi
Faculty of Civil Engineering, Universiti Teknologi MARA, 40450 Shah Alam, Selangor, Malaysia
e-mail: raizam@uitm.edu.my

S. W. Lee
Faculty of Civil Engineering, Universiti Teknologi MARA, Johor Pasir Gudang Campus, 81750 Masai, Johor, Malaysia

S. S. Emamian et al. (eds.), *ICSDEMS 2019*,
https://doi.org/10.1007/978-981-15-3765-3_24

In order to control cracks on concrete structures, load-bearing capacity was introduced, and its application can be detected on many construction fields like concrete footing, bridge pedestals, anchorage zone, bridge bearing and others [4]. The load-bearing strength was first studied by [5] followed by [6–8]. Many researchers found that there was a formation of an inverted pyramid under the loading bearing plate and expressed a formulation for concrete bearing capacity based on that observation. Shelson derived an empirical formula to get the bearing strength of concrete [7], as in Eq. (1).

$$f'_{cc} = f'_c\left(\sqrt[3]{\frac{A_2}{A_1}}\right) \tag{1}$$

where f'_{cc} represents the ultimate bearing strength of concrete, $f_c{}'$ as concrete strength, A_2 denotes as a gross area of the concrete foundation and A_1 as concrete bearing area or bearing plate area. Niyogi has carried out several testing on concrete and has demonstrated that the ratio (length/width) that was greater than two (2), to make the bearing strength, was almost constant for blocks concrete [9]. The researcher proposes an empirical equation for the bearing strength of concrete blocks under concentric compressive load as written in Eq. (2).

$$\frac{f_b}{f'_c} = 0.84\left(\frac{\alpha}{\alpha'}\right) + 0.23 \tag{2}$$

In which, f_b represents the concrete bearing strength, a represents the dimension of concrete blocks and a' denotes the steel plate dimension. The enveloping concrete gives an effect to the higher compressive strength of concrete at bearing area as stated by [1, 10]. Thus, in view of this and keeping in mind the usefulness of bearing strength to the design of structural members, experimental investigations were carried out to estimate the optimum bearing capacity of concrete blocks axially loaded via different sizes of steel plates and observe the possible failure modes of the concrete.

2 Method

A series of experimental investigations have been performed to calculate the bearing strength of plain and reinforced concrete blocks and observe its possible mode of failure. The test was carried out subjected to concentric compression load. Water–cement ratio of 0.46 was chosen to get the maximum strength and durability. In this research, square bearing steel plates (SQ) with dimensions of 50 mm × 50 mm, 100 mm × 100 mm, and 150 mm × 150 mm in cross-section and 10 mm thickness were used. Ordinary Portland Cement was used due to its advantages of having a medium rate of hardening and providing enough level of strength and durability. Natural granular material was used as fine aggregate (0.0625–2 mm) and the coarse

Table 1 Concrete mix proportions

Water (kg/m³)	Cement (kg/m³)	Fine aggregate (kg/m³)	Coarse aggregate (kg/m³)	
			10 mm	20 mm
180	390	525	426	852

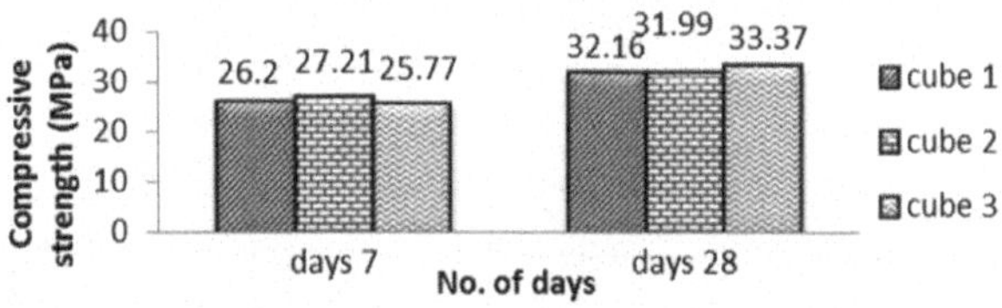

Fig. 1 Mechanical properties of the concrete mix

aggregates were sieved to the specific size in the range between 10 and 20 mm. Reinforced concrete (concrete consist of reinforcement bar embedded in the concrete) was chosen as the concrete block specimens, while plain concrete acting as a control specimen. All concrete block specimens with the characteristic strength of 30 N/mm^2 were cast from a concrete mix as depicted in Table 1. Then, the concrete was poured into the prepared mould.

Twelve (12) numbers of concrete blocks (200 × 200 mm) including plain concrete (PC) and reinforced concrete (RC) blocks were prepared. From a batch of concrete, six 100 mm × 100 mm concrete cubes were cast as control specimens. The ingredients of the concrete were mixed up until uniform consistency was achieved. The concrete cubes were tested up to the failure for their compressive strength after being stored in a curing tank for 7 and 28 days to estimate the mechanical properties of the mix as depicted in Fig. 1.

The loading rate is set for 6.8 kN/s based on the pace rate limit as specified in ASTM C29. A total of twelve (12) numbers of concrete block specimens were cast and tested up to failure subjected to concentric compression load after 28 days of achieving its designated age. Detailed configurations of the concrete block specimens are shown in Fig. 2.

2.1 Experimental Details and Test Set-up

Load cell and displacement transducers were connected to a digital data logging system to measure load-bearing strengths and displacement values. Every single test of specimen was subjected to a constant axial load (2300 kN). The specimens were loaded without shock and continuously until failure. The test set-up for the specimens is shown in Fig. 3. A linear variable differential transformer (LVDT) was used to determine cross-head displacement under the applied bearing load. The ultimate load and any crack occurred on the concrete block specimens were recorded and observed.

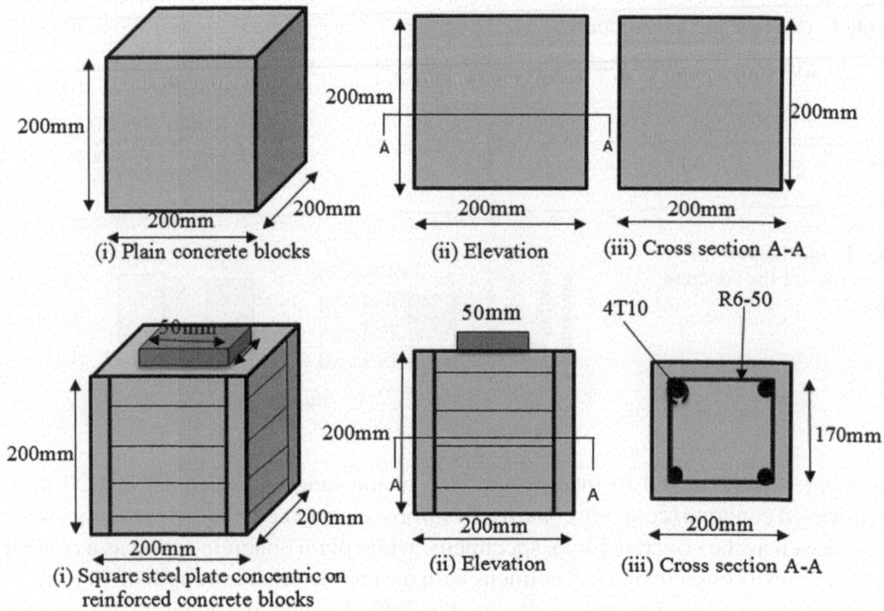

Fig. 2 Details notation of plain and reinforced concrete block specimens

Fig. 3 Test set-up and position of LVDT

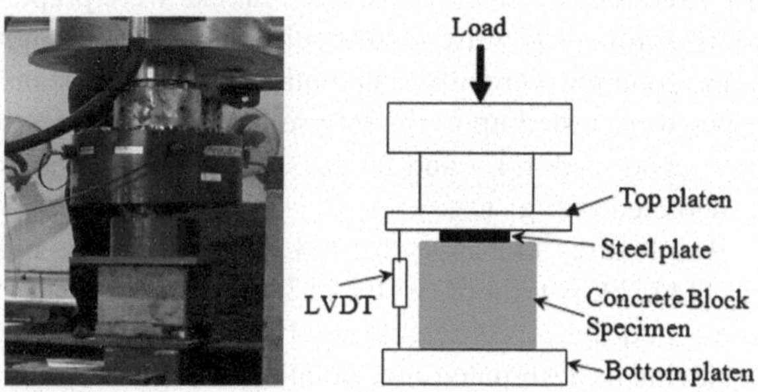

3 Bearing Strength Behaviour Under Concentric Load

The maximum load capacity and displacement values were attained using a data acquisition system, recorded by the computer. The average values of load-carrying capacity and crack patterns were used to estimate the ultimate bearing capacities and the failure modes for concrete block specimens loaded through different sizes of square steel plates, respectively. As can be seen in Table 2, the average bearing strength for the plain concrete without steel plate (PC) was recorded as 0.023 kN/mm^2 with the average displacement of 10.75 mm. Reinforced concrete block specimens loaded through square steel plate denoted as RC-SQ50 gave the average maximum value of bearing capacity as 0.154 kN/mm^2 with 6.29 mm displacement. The average

Table 2 Results of bearing strength test

Specimen designation	Average load capacity (kN)	Average displacement (mm)	Average bearing strength (kN/mm^2)
PC	920.00	10.75	0.023
RC-SQ50	384.81	6.29	0.154
RC-SQ100	662.67	8.60	0.066
RC-SQ150	885.50	10.13	0.039

Table 3 Comparison between the experimental test and earlier studies

Specimens designation	f_b (exp) (kN/mm^2)	$f_{b(N)}$ (Nayogi, 1973) (kN/mm^2)	$f_{b(S)}$ [7] (kN/mm^2)	Ratio $f_b/f_{b(N)}$	Ratio $f_b/f_{b(S)}$
RC-SQ50	0.154	0.410	0.080	0.38	1.93
RC-SQ100	0.066	0.110	0.050	0.60	1.32
RC-SQ150	0.039	0.052	0.040	0.75	0.97

bearing strength and average displacement values for RC-SQ100 was recorded as 0.066 kN/mm^2 and 8.60 mm, while for RC-SQ150, its average maximum value of load-bearing strength was recorded as 0.039 kN/mm^2 with 10.13 mm displacement. The experimental result for each type of specimen and its comparison with earlier studies are shown in Tables 2 and 3.

Based on the experimental results, concrete block specimens with the steel plate dimensions of 50 mm x 50 mm gave higher values of bearing strength compared with the steel plate with the dimensions of 150 mm × 150 mm and 100 mm × 100 mm. Experimental results demonstrate that the value of deformation also become larger as greater cross-sectional area of the steel plate was used. The maximum bearing strength of concrete can be increased by increasing the ratio of unloaded to the loaded area. The load that implies the concrete specimen was transferred through a square steel plate into the concrete block with a larger contact area [2]. This indicates that the increase in the surrounding area corresponds with value in terms of load-carrying capacity before the concrete specimen reaches its limiting value [11].

3.1 Mode of Failures

The type of failure modes that have been detected during the experimental testing is in the forms of localized damage, which can be noticed at the outer edge of the contact area. It has been found that these observations, especially for unreinforced concrete blocks are comparable with earlier studies [2, 12] where the vertical cracks, non-explosive cracks, and inverse pyramid failures have been detected at the outer edge of contact area, as shown in Figs. 4, 5, 6 and 7.

(i) Vertical crack (ii) Non-explosive crack (iii) inverse pyramid shape

Fig. 4 Modes of failures for unreinforced concrete blocks

(i) Vertical crack (ii) Non-explosive crack (iii) Brittle crack (iv) Radial crack

Fig. 5 Modes of failures for reinforced concrete blocks concentrically loaded through 50 mm × 50 mm square steel plates

(i) Vertical crack (ii) Split crack (iii) Edges crack (iv) Radial Crack

Fig. 6 Modes of failures for reinforced concrete blocks concentrically loaded through 100 mm × 100 mm square steel plates

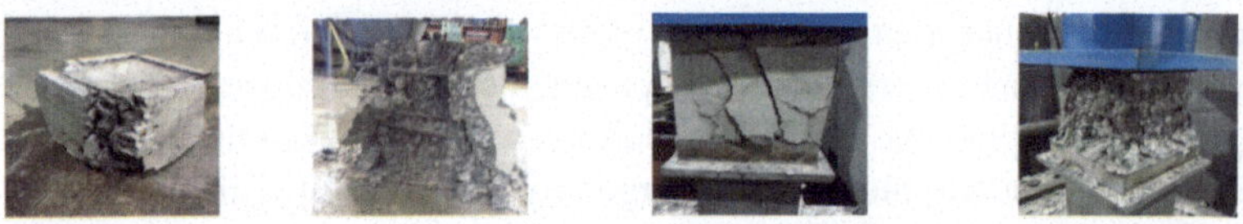

(i) Wedge crack (ii) Brittle crack (iii) Vertical crack (iv) Non-explosive crack

Fig. 7 Modes of failures for reinforced concrete blocks concentrically loaded through 150 mm × 150 mm square steel plates

As can be seen in Fig. 4, the types of failure mode for plain concrete block without steel plate on top of it are vertical cracks, non-explosive crack, and inverse pyramid failure. These types of failure occur when the maximum tensile stress at the top of the concrete exceeds the tensile strength of the concrete itself. In addition, when the reinforced concrete block was loaded through the steel plate size of 50 mm × 50 mm, the failure that could be observed was vertical cracks. While radial and punching cracks were noticed when indentations within ranges of 3–6 mm were spotted on the concrete block specimen as shown in Fig. 5iv. When the plate dimension of 100 mm

× 100 mm was used, the mode of failure that could be observed was splitting crack. Large splitting wedge punched out from beneath the steel bearing plate as can be observed in Fig. 6ii, iii. However, as can be seen in Fig. 7iv, the concrete block that was loaded through a steel plate with the dimensions of 150 mm × 150 mm shows the non-explosive and brittle modes of failure.

4 Conclusions

The conclusions of the present study were drawn as follows:

(1) Concrete block specimens loaded through smaller sizes of steel plate gave lower values of deformations and higher values of bearing strengths.
(2) For all reinforced concrete block specimens, when the dimension of the steel plate used increases, the ratio of the concrete specimen area to the area of bearing plate tends to decrease, thus resulting in the decrement value of ultimate bearing strength.
(3) High-stress concentration at the sharp edge of the square steel plate contributes to the lower values of bearing strength of concrete block specimens.

Acknowledgements The authors would like to acknowledge the Ministry of Higher Education under Research Acculturation Grant Scheme (RAGS) (RAGS/1/2015/TK01/UITM/03/1) for providing financial support to perform this study. Authors would like to thank especially Miss Nurul Adilah binti Mohd who directly was involved in this research. Also, to all laboratory staff in Concrete, Fabrication, and Heavy Structures Laboratories, Faculty of Civil Engineering, Universiti Teknologi MARA (UiTM), Shah Alam for their strong support throughout this research works.

References

1. Escobar-Sandoval, D., Andren, S. W., & Gary, F. D. (2006). Concentrically loaded circular steel plates bearing on plain concrete. *Journal of Structure Engineering, 132*(11), 1784–1792.
2. Mohd Raizamzamani, M. Z. & Norrul, Azmi, Y. (2017). Experimental study on bearing strength of concrete blocks under concentric compression load. *Pertanika Journal of Science And Technology*, **25**(S): 67–76.
3. Rodolfo, B., Carin, L. R. & Santos, J. T. (2014). Bearing strength of confined concrete. *ACI Structural Journal*, 1317–1327.
4. Mohd Raizamzamani, M. Z., Yahya, N. A., Wee, L. S., Lian, O. C. & Yunus, B. M. (2019). Experimental study on the bearing capacity of concrete blocks loaded through square and circular steel plates. *IOP Conference Series: Materials Science and Engineering*, vol. 513, pp. 1–10.
5. Baushinger, J. (1876). Tests with blocks of natural stone. *Mechanisch und Technischen Laboratorium der Kgl. Technischen Hochschule*, Munich, 6, 13.
6. Meyerhof, G. G. (1953). The bearing capacity of concrete and rock. *Magazine of Concrete Research*, 107–116.

7. Shelson, W. (1957). Bearing capacity of concrete. *Journal of the American Concrete Institute, 54*(5), 405–414.
8. DeWolf, J. T. (1978). Axially Loaded Column Base Plates. *Journal of the Structural Division ASCE, 104*(ST4), 781–794.
9. Niyogi, S. (1973). Bearing Strength of Concrete-Geometric Variation. Niyogi, S. (1973). Bearing Strength of Concrete-Geometric Variation. *J. Struct. Div., ASCE, 99(ST7),* 1471–1490.
10. Ahmed, T., Burley, E., & Rigden, S. (1998). Bearing capacity of plain and reinforced concrete loaded over a limited area. *ACI Structural Journal*, 330–340.
11. Rao, C. K., & Swamy, R. N. (1974). Bearing strength of steel fibre reinforced concrete. *Building Science, 9*(4), 263–268.
12. Zhou, W., Hu, H., & Zheng, W. (2013). Bearing capacity of reactive powder concrete reinforced by steel fibers. *Construction and Building Materials*, 1179–1186.

Bearing Strength of Concrete Blocks with External Wrapping of Carbon Fibre Reinforced Plastic (CFRP)

Norrul Azmi Yahya, Norliyati Mohd Amin, Mohd Raizamzamani Md Zain, Oh Chai Lian, and Lee Siong Wee

Abstract Large heavy force transfers over a limited area of concrete surface can cause high concentrated stresses due to contact interaction of steel bearing plate on concrete surfaces has been highlighted in designing the bearing strength of concrete. For concrete bearing, the bearing capacity increases due to confining pressure provided by a steel bearing plate. The bearing capacity of concrete can be greater than its compression strength depends on the concrete to steel area (Ac/As) ratio. The steel bearing plate acted as load dispersing plate in the load transfer members. The present study examines the confinement effect for CFRP wrapping to load-bearing capacity of concrete. Six numbers of concrete cube blocks of 150 mm^3 were loaded through the steel bearing plate with sizing of 100 mm^2 up to failure level under vertical load. The significant effect of external reinforcement of Carbon Fibre Reinforced Polymer (CFRP) to concrete bearing was examined. The experimental results indicate that the structural performance of concrete bearing wrapping using CFRP provides additional lateral confinement can increase bearing capacity of concrete.

Keywords Bearing strength of concrete · External wrapped concrete · Carbon Fibre Reinforced Pastic (CFRP) · Lateral confinement effect

N. A. Yahya (✉)
Centre of Postgraduate Studies, Faculty of Civil Engineering, Universiti Teknologi MARA, 40450 Shah Alam, Selangor, Malaysia
e-mail: azmi_216@yahoo.com

N. M. Amin
Head of Centre of Studies (Structure), Faculty of Civil Engineering, Universiti Teknologi MARA, 40450 Shah Alam, Selangor, Malaysia

M. R. M. Zain · O. C. Lian
Faculty of Civil Engineering, Universiti Teknologi MARA, 40450 Shah Alam, Selangor, Malaysia

L. S. Wee
Faculty of Civil Engineering, Universiti Teknologi MARA, 81750 Pasir Gudang, Johor, Malaysia

S. S. Emamian et al. (eds.), *ICSDEMS 2019*,
https://doi.org/10.1007/978-981-15-3765-3_25

1 Introduction

The bearing strength of concrete is very important to provide structural stability at the support when heavily loaded through a concentrated area of concrete surface. Bearing strength is vital in concrete beds, bridges pedestals, post-tension members, concrete corbels, concrete anchorage, and column—footing connection design. These structure supports are exposed to experience high concentrated loads that may affect the structural integrity of the structure. The large concentrated loads acted on the concrete surface can lead to uneven load distribution of concrete surfaces.

The heavy concentrated load on concrete bearing surface becomes vital as confining pressure was increased in the enveloped area [1–3]. Generally, the bearing strength of concrete depends on the ratio of concrete area to bearing area and the compressive strength of concrete. However, heavy concentrated load can be exceeded the bearing capacity of concrete which susceptible to localized failure such as edge cracking or spalling of concrete beds [4, 5]. According to Australian Standard, the bearing strength of concrete [6] can be taken as:

$$\frac{f_b}{f_c'} = 0.8\left(\frac{A_2}{A_1}\right)^{0.5} \leq 2.0$$

Whereas, f_b is the bearing strength of concrete, f'_c is the compressive strength of concrete, A_1 is the area of bearing plate, and A_2 is the total surface of the concrete.

External reinforcement in structural strengthening is currently available to restore structural integrity. The external wrapping, such as CFRP wrapping, widely used in column and bridge girder strengthening for buildings and bridges [7–9]. It was proven that the structural performance of wrapped structure members was able to restore the load capacity and structural performance.

The current research examines the external reinforcement technique to improve structural performance of concrete bearing using the Carbon Fibre Reinforced Plastic (CFRP) wrapping. It was noted that the existing formulation for concrete bearing does not include the effect of external confinement directly [10, 11, 12]. There are also limited fundamental studies on Carbon Fibre Reinforced Polymer (CFRP) wrapping for concrete bearing. Therefore, this study is carried out to determine the structural performance of wrapped concrete bearing using CFRP.

2 Methodology

2.1 Compression Test

The concrete was cured in water with temperature 27 °C in the curing tank and tested for 28 days. The test was carried out using the Compression Testing Machine (CTM-1000kN). The concrete block volume is 150 mm^3. The consistent loading rate is used for 9.2 kN/s.

Fig. 1 Concrete block wrapped with CFRP

2.2 *External CFRP Wrapping*

Carbon polymer reinforced fibre (CFRP) sheet acts as an external reinforcement ideally to improve the confinement effect of concrete. The surface of the concrete is clean from dust before wrapping. The concrete surface was a blast with water to improve its roughness. The surface roughness can increase the effectiveness of the bonding between the CFRP sheet and the concrete surface. For CFRP wrapping, epoxy was used as a binder. The Sikadur 30 epoxy acted as an adhesive element to bind the CFRP sheet and the concrete surface. The epoxy also acted as filler especially at the uneven surfaces in order to get better results in the load capacity. For the current research, only one layer of CFRP sheet was used and it was fully wrapped on the outside surface of the block as shown in Fig. 1.

2.3 *Bearing Test*

Six concrete blocks with a dimension of 150 mm^3 were tested using Compression Testing Machine (CMT-1000kN) with 1000 kN capacity (see Fig. 2). There are two types of concrete bearing, i.e. wrapped with CFRP and unwrapped. The bearing steel plate is 100 mm^2 with 10 mm thickness. The rubber pad placed under the steel bearing plate to avoid uneven contact surface. The steel bearing plate was placed directly at the centre of concrete surface. In order to improve the distribution of the loaded area, the concrete surface must be free from any substance.

Fig. 2 CFRP concrete wrapped loaded by steel bearing plate

Table 1 Results of compressive strength

Sample	Time (Days)	Compressive strength (N/mm^2)	Average
1	28	41.30	41.62
2		41.01	
3		42.55	

3 Results and Discussions

3.1 Compression Test

Compressive strength is also recognized as the maximum capacity loading of concrete which is able to withstand the compressive load. In the present study, the concrete addictive SikaSet Low Shrinkage Accelerator has been added during mixing the mixture. The loading rate for pace was consistently set at 6.8 kN/s. The results of compressive test are tabulated in Table 1.

Table 1 shows the compressive strength for three samples after 28 days are 41.30, 41.01, and 42.55 MPa. The average of compressive strength is 41.62 MPa.

3.2 Bearing Strength

In previous studies, there are many factors affecting the bearing strength of concrete. One of the factors is the external wrapping of CFRP. The concrete block was centrally loaded from the top surface of concrete using a steel bearing plate of 100 mm^2. The linear variable displacement transducer (LVDT) apparatus also was used to monitor the displacement of the concrete block during the testing. The results of the bearing test for a concrete block of 150 mm^3 were tabulated in Table 2.

Table 2 shows the ultimate load for unwrapped and CFRP wrapped concrete blocks. The surface area of concrete block and steel plates were 22,500 mm^2 and 10,000 mm^2, respectively. The area surface of concrete block (A_2) to area of square plate (A_2/A_1) ratio is 2.25. The bearing strength (f_b) is calculated based on the ultimate load (kN) divided by the contact area. The compressive strength of concrete at day 28 days was taken from compressive test results. The confinement effect is a dimensionless unit which was calculated based on the ratio bearing strength to compression strength. The average bearing strength for wrapped and unwrapped is 54.70 N/mm^2 and 49.76 N/mm^2, respectively. This indicates that the concrete block wrapped with CFRP has greater bearing strength compare to normal concrete bearing with no wrapping. The result is also similar to [9] which proven that CFRP can provide additional lateral confinement for concrete that can ultimate bearing strength of concrete.

Figure 3 shows the load versus displacement graph for each wrapped and unwrapped concrete block. The CFRP wrapped concrete blocks tend to have higher load resistivity compare to unwrapped concrete blocks. The CFRP wrapped concrete blocks also performed better as stiffness increased as shown in the graph.

Figure 4a, b illustrate that the concrete blocks wrapped with CFRP have high confinement effect and high bearing strength for 54.70 N/mm^2 as compare to unwrapped (49.76 N/mm^2). The confinement effect of concrete blocks that were unwrapped and wrapped were 1.20 and 1.31, respectively. This proved that Carbon Fibre Reinforced Polymer (CFRP) wrapped at the outer surface of the concrete improved the confinement effect from 20% up to 31%.

Table 2 Bearing Strength of square concrete block at 28 days

Sample	Types	f_{ult} (kN)	A_2/A_1	f_b (N/mm^2)	Average	f_b/f_c	Average
1	No CFRP	457.7	2.25	45.77	49.76	1.10	1.20
2		510.0		51.00		1.22	
3		525.2		52.52		1.26	
4	With CFRP	548.5		54.85	54.70	1.31	1.31
5		550.5		55.05		1.32	
6		542.0		54.20		1.30	

Note A_1 = Area of square plate with thickness of 10 mm (10,000 mm^2), A_2 = Surface area of square concrete block (22,500 mm^2), f_{ult} = Ultimate load, f_b = Bearing stress of confined circular concrete (circular plate), and f_c = Compressive strength of unconfined concrete block

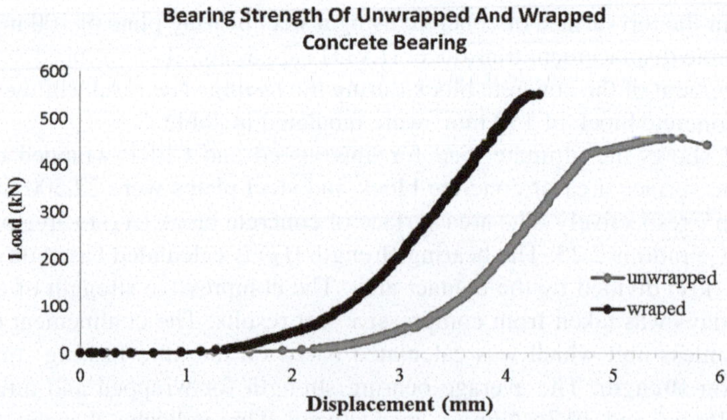

Fig. 3 Bearing strength of unwrapped and wrapped concrete bearing graph

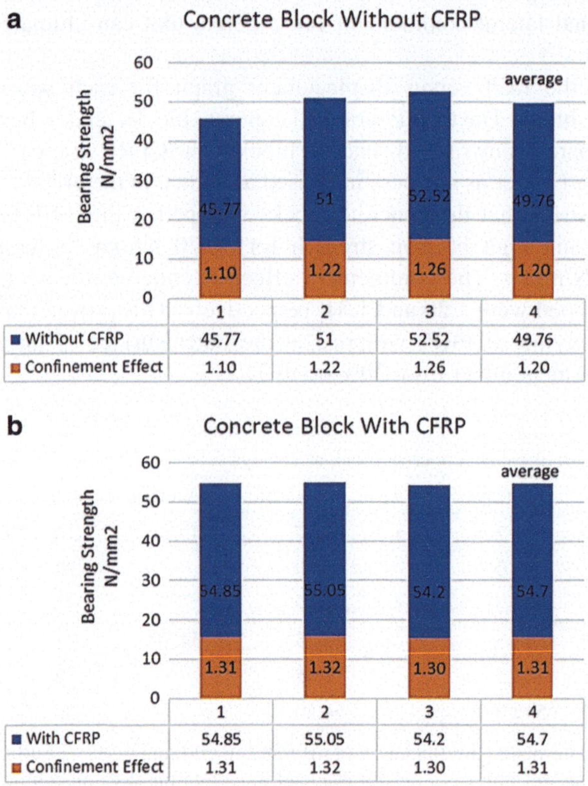

Fig. 4 **a** Bearing strength and confinement effect of unwrapped concrete blocks. **b** Bearing strength and confinement effect of CFRP wrapped concrete blocks

3.3 Failure Modes

The failure mode of concrete bearing can be identified through a cracking pattern that appears on the concrete surface after reaching the maximum load. The failure modes of bearing strength can be observed and the cracking pattern which usually started at the corner of steel bearing plate as observed in Fig. 5a, b.

Table 3 shows the failure mode of concrete block with and without CFRP wrapping. For the unwrapped concrete block, the concrete surface experience brittle fractures and vertical crack patterns under the bearing steel plate. The cracking of the concrete bearing was observed starting from the corner of the square bearing plate in which cracking pattern appears and spreads on the top of the surface cause splitting failure of the concrete. For the CFRP wrapped block, the failure initiates with vertical cracking and spread vertically in different angles as shown in Fig. 5b. The maximum bearing load was recorded shortly after cracking occurred and spread throughout the entire square concrete block. The majority of the cracks started from the edge corner of steel bearing plate. It was noted that the level of damage of concrete block wrapped with CFRP is less than unwrapped concrete block which confirming the

Fig. 5 **a** Failure mode for unwrapped concrete block. **b** Failure mode for CFRP wrapped block

Table 3 Different types of mode failure

CFRP on concrete blocks	Types of mode failure
Unwrapped	Splitting crack, brittle fracture and vertical crack
Wrapped	Edge crack

CFRP wrapped blocks has experienced higher confinement effect as compared to unwrapped CFRP concrete blocks.

4 Conclusions

The presence of CFRP on concrete block increases the confinement effect of bearing strength of concrete. The following final conclusion can be drawn from the research:

1. Bearing strength of concrete has been improved due to an increase of lateral confinement provided by Carbon Fibre Reinforced Polymer (CFRP) wrapping.
2. The CFRP wrapped concrete blocks exhibit greater bearing capacity than unwrapped concrete. It was found that CFRP also suitable material for external wrapping.
3. The confinement effect (f_b/f_c) of concrete block wrapped with CFRP is slightly higher compared to unwrapped concrete blocks which explain the concrete wrapped that posed a significant increase of bearing strength. The presence of CFRP on concrete blocks will increase the bearing strength up to 31%.
4. The failure mode for wrapped and unwrapped CFRP blocks posted significant types of failure modes that were observed up to failure. The unwrapped concrete blocks experienced major failure through formation of splitting, vertical crack, and brittle fracture lead to total failure. However, CFRP wrapped blocks experience lesser level of damage.
5. The crack pattern or localized failure is critical due to penetration of loaded plate that relation to sharp edge contact stress. This was confirmed as the formation of cracking was initiated at the sharp edge of steel bearing plate.

Acknowledgements The authors gratefully acknowledge the financial supports by Universiti Teknologi MARA (UiTM) under BESTARI PERDANA Research Grant 600-IRMI/PERDANA 5/3 BESTARI (060/2018). Authors would also like to thank especially undergraduate research student Mr. Muhammad Nadzmi Mat Salam who was directly involved in this experimental works. Also, to all laboratory staff in Concrete, Fabrication and Heavy Structures Laboratories, Faculty of Civil Engineering, Universiti Teknologi MARA (UiTM), Shah Alam for their support.

References

1. Md Zain, M. R., & Yahya, N. A. (2017). Experimental study on bearing strength of concrete blocks under concentric compression load. *Pertanika Journal Science Technology, 25*(S): 67–76.
2. Ince, R., & Arici, E. (2004). Size effect in bearing strength of concrete cubes 18: 603–609.
3. Program, E. (2016). Study on effect of bearing strength on rectangular and square pedestals with square bearing plates subjected to concentric loading 3(4): 173–179.

4. Wang, W., Tang, Z., Li, Z., Ma, H. et al. (2016). Bearing capacities of different-diameter concrete filled steel tubes under axial compression. *Advances in Materials Science and Engineering,* 2016.
5. Yahya, N. A. (2017). Strategies for mitigation of the failure of concrete pedestals supporting bridge girder bearings. Ph.D. Thesis Queensland University of Technology.
6. AS3600. (2009). Australia standard for concrete structures; Standards Australia Limited.
7. Li, G. (2006). Experimental study of FRP confined concrete cylinders 28: 1001–1008.
8. Scheffers, C. A. (2010). Bearing strength of CFRP confined concrete 6(1): 1–3.
9. Ravindrarajah, R. S., & Reinaldy, R. (2010). Bearing strength of CFRP confined concrete.
10. Yahya, N. A., Dhanasekar, M., Abd Hamid, H. Localised failure mechanism of concrete pedestals under bridge bearing. Australian Journal of Civil Engineering: 1–12.
11. Yahya, N. A., Dhanasekar, M., Md Zain, M. R., Oh, C. L., Lee, S. W. (2019). Numerical modelling of concrete bearing strength for different heights of concrete blocks. IOP Conference Series: Materials Science and Engineering 513: 012031.
12. Yahya, N. A., Dhanasekar, M., Md Zain, M. R., Oh, C. L., Lee, S. W. (2018). Bearing strength of concrete for difference heights of concrete blocks. IOP Conference Series: Materials Science and Engineering 431: 072011.

Mechanical Properties of Engineered Cementitious Composite (ECC): An Overview

Nurmazidah Abdullah Zawawi, Chai Lian Oh, Siong Wee Lee, Mohd Raizamzamani Md Zain, and Norrul Azmi Yahya

Abstract This paper presents an overview of the mechanical properties of Engineered Cementitious Composites (ECC) or more popularly known as bendable concrete. Effects of various mineral admixtures as cement replacements such as ground granulated blast-furnace slag (GGBS), fly ash (FA) and glass powder (GP) on the mechanical properties of ECC specifically tensile, compressive and flexural properties are discussed. Mechanical properties of ECC with the inclusion of steel (SE) fiber, polypropylene (PP) fiber, polyethylene (PE) fiber and polyvinyl alcohol (PVA) fiber in the past experimental works are also presented. With the comparison between previous studies presented in this paper, the authors hope to identify the limitation of the current ECC mixture, so that improvements can be done for future studies.

Engineered cementitious composite · Cement replacement · Flexural and tensile strength

1 Introduction

Engineered Cementitious Composites (ECC) was originally developed by Li at the University of Michigan in 1993. ECC is characterized by excellent strain capacity under uniaxial tension, average fine crack width and with moderate fiber volume fraction (typically 2%). Unlike conventional concrete that demonstrates strain-softening after first cracks, ECC exhibits tensile strain-hardening behaviour. High ductility of ECC is due to the interaction between the fiber, ECC matrix and fiber-matrix interface [1]. Additionally, ECC demonstrates high tensile strength and good durability due to self-controlled cracks. This phenomenon can be seen through a series of formations

N. A. Zawawi · C. L. Oh (✉) · M. R. M. Zain · N. A. Yahya
Faculty of Civil Engineering, Universiti Teknologi MARA, 40450 Shah Alam, Selangor Darul Ehsan, Malaysia
e-mail: chailian@uitm.edu.my

S. W. Lee
Faculty of Civil Engineering, Universiti Teknologi MARA Johor, Pasir Gudang Campus, 81750 Bandar Seri Alam, Masai, Johor, Malaysia

S. S. Emamian et al. (eds.), *ICSDEMS 2019*,
https://doi.org/10.1007/978-981-15-3765-3_26

and spreads of a small number of cracks within the material after initial loading until the widening of cracks were maintained at an average width of about 60 μm, resulting from micro-mechanical tailoring of ECC [2]. Application of only fines in ECC mixture in order to obtain better strain capacity and fine crack width has resulted in the usage of higher cement content compared to ordinary concrete. Due to environmental and economic reasons, there is a growing trend to use industrial wastes or by-products as supplementary materials in the production of composite cement. Therefore, popular mineral admixtures such as fly ash (FA), silica fume (SF) and ground granulated blast furnace slag (GGBS) have been considered to replace the cement in the production of green ECC which can promote the global sustainable development.

This paper presents an overview of the important constituents in ECC and the pioneer studies on the mechanical properties particularly compression, flexural and tensile behaviours of ECC.

2 Constituents in ECC

Overview of the important constituents such as fiber and mineral admixtures employed to produce ECC is presented in this section. The use of fiber in ECC significantly improves the tensile ductility properties of ECC with a certain volume fraction. Fibers such as steel fiber (SE), polypropylene fiber (PP), polyethylene fiber (PE) and polyvinyl alcohol (PVA) fiber are commonly used in ECC mixes. Table 1 shows the properties of fiber used in previous investigation works. The application of the fibers can improve strength and strain capabilities as well as energy absorption of ECC [3]. This further reduces the damage of the ECC structure during an impact loading. High-modulus fibers such as SE, glass fiber and carbon fiber increase the bulk strength and toughness of the material, however, their intrinsic brittle behaviour does not allow for ductility or strain hardening in ECC [4]. On the contrary, low-modulus fibers such as PVA, PP and PE fibers significantly improve the ductility of

Table 1 Properties of fiber

Authors	Type	Nominal strength (MPa)	Diameter (μm)	Length (mm)	Young modulus (GPa)	Density (kg/m^3)
Şahmaran & Li [2]	PVA	1600	39	12	40	–
Maalej et al. [4]	PE	2610	39	13	66	–
Lee et al. [11]	PP	–	19.5	6	–	910
Pourfalah [13]	SE	3000	150	6, 13	210	7800

Note Dash indicates no information provided from the respective article the concrete mix at the same time reduce cracking. It has been found that the mechanical properties of ECC such as drying shrinkage, strain capability, compressive and flexural strength can be improved with the application of both hybrid low—and high-modulus fibers at a certain volume [4]. PVA fiber was recommended as one of the most suitable polymeric fibers for ECC materials [5]. PVA fiber exhibits some improved mechanical properties such as tensile strength and Young's modulus compared to PP and polyolefin fibers [6].

Incorporation of mineral admixtures such as fly ash (FA), and ground granulated blast furnace slag (GGBS) in the ECC production has resulted in excellent properties such as demonstration of improved workability, cracking and desirable long-term mechanical properties especially durability [7]. FA and GGBS are the by-products of the combustion of powdered coal in power plants and iron blast furnace, respectively. A recent development in the production of ECC has incorporated FA as a partial replacement for portland cement [8–10]. The addition of FA to ECC improved robustness of tensile ductility while retaining a long-term tensile strain of approximately 3% and reducing the crack width (i.e. less than 60 μm) [9]. On the other hand, mechanical properties of ECC with high GGBS to cement replacement ratio of up to 50 to 70% were investigated [11, 12]. It is found that there was a reduction in compressive strength for ECC mix with 70% GGBS replacing cement compared to the mixes with lower percentages of GGBS.

3 Mechanical Behaviour of ECC

Investigation works to obtain mechanical behaviour of ECC particularly compressive, tensile and flexural properties are reviewed in this section and are summarized in Table 2.

Generally, the compressive strength of ECC was obtained through uniaxial compression test on a series of cylinder or cube specimens at 28 days of curing. Investigation on the compression strength of ECC by replacing the cement with 50%, 60% and 70% of GGBS was conducted [12]. Highest compression strength was observed in ECC mix with 60% GGBS. It was also found in the same study that there is no significant effect on the mechanical properties of ECC with different PP fiber contents (i.e. 1.5%, 2.0% and 2.5% in volume fraction). On the other hand, the compressive strength development of the ECCs with different percentages of FA and GP at 7, 28, 56, 90 and 120 days were reported [7]. Highest compressive strength was reported for ECC mix containing 25% GP replacing FA after 28 days of curing compared to mixes with 50%, 70% and 100% replacements. High proportion of GP with respect to FA may give rise to the effect of pozzolanic reaction and cement hydration. Another experimental work also investigated compressive strength of ECC mix with as high as 70% replacement cement but with only FA [8]. A study on compressive strength of ECC specimens after exposure to high temperatures such as 200 °C, 400 °C and 600 °C was also carried out [13]. At ECC mix with 60% FA replacement of cement, the compressive strength for specimens that were exposed

Table 2 Mechanical properties of ECC

Mixture	Previous investigation works					
	Lee et al. [12]	Wang & Li [9]	Şahmaran et al. [8]	Pourfalah [13]	Siad [7]	Pourfalah [13]
GGBS/cement	1.5	–	–	–	–	–
Fly Ash/cement	–	2.2	1.2	1.8	–	1.8
Sand/binder	0.2	0.69	0.36	0.2	0.41	0.2
Glass powder/binder	–	–	–	–	0.14	–
Water/binder	0.27	0.28	0.27	0.28	0.27	0.28
Fiber in volume friction (%)	2.0					
Type of fiber	PP	PVA				Steel
Type of sand	River	Silica sand				
Compressive strength (MPa)	21.4	35.0	62.5	38.8	62	40.1
Tensile strength (MPa)	3.43	4.8	5.14	3.81	–	4.3
Tensile strain capacity (%)	1.56	3.9	2.73	3.5	–	3
Flexural (MPa)	–	–	–	–	10.45	–

to 200 °C, 400 °C and 600 °C decreased 6%, 10% and 25%, respectively compared to their counterpart under ambient temperature (20 °C).

On the other hand, tensile strain capacity and crack control are outstanding characteristics for ECC. Many have proposed new ECC mixes and investigated the tensile strength of the material. Dog-bone shape specimen is generally used in the uniaxial tensile test. Wang and Li (2007) claim that the increment of FA content in ECC can improve the tensile strain of the material. Tendency of FA in reducing chemical bonding between PVA fiber and ECC matrix and at the same time increasing the interface frictional bond was highlighted. Additionally, the significant reduction of crack width was found at the increment of FA content at all ages. This is likely associated with the higher fiber/matrix interface frictional bond when there is an increase of FA content [9, 10]. Higher tensile strength can be achieved by reducing water binder ratio which also enhances matrix strength. The conclusion that higher ductility ($3.6 \pm 0.3\%$ for the tensile strain capacity) and tensile strength (5.2 ± 0.2 MPa for the tensile strength) for the ECC mix containing GGBS compared to the ECC mix without GGBS has been made [14]. It is noted that tensile behaviour of ECC is sensitive to the change of temperature. ECC material that is subjected to high temperature demonstrated a reduction in strain hardening capacity and lower characteristic tensile strength which may eventually lead to brittle failure [15]. The obvious reduction in tensile strength can be seen especially after the deterioration of fibers at temperature 150 °C. The polymeric fibers of ECC (i.e. PVA fiber) can decrease the degree of spalling and explosion spalling at high temperatures [8].

Flexural behaviour in terms of load-deformation of ECC is important as the deflection can reflect the ductility of the material. Investigation work on the performance of ECC mixes with different percentages of GGBS and FA replacing cement (i.e. ratio GGBS:FA 70:0, 60:10, 50:20 and 40:30) was conducted [16]. Higher load-carrying capacity was observed in ECC mix with a combination of GGBS and FA compared to ECC mix with only FA (specifically 0% GGBS and 70% FA replacing cement) at age of 3 to 90 days [16]. Obvious reduction in deflection for ECC specimens was observed in ECC mix incorporating as high as 70% combination of mineral admixtures replacing cement (i.e. GGBS and FA) at curing age of 90 days (with approximately 12 mm) compared to curing age of 3 days (more than 40 mm). Alternatively, ECC mixtures indicate a probable enhancement in the fiber-matrix frictional bond strength with GP content. The overall decrease in ECC ductility with GP replacement can also be associated with the higher alkali content of GP which may enhance bond strength and fiber-matrix chemical bond. Moreover, the FA and GP particles have very different morphologies, which can influence fiber-matrix interface properties. Unlike the spherical morphology of FA particles, glass powders consist mainly of fine angular particles [5]. On the one hand, fiber strongly influences the flexural behaviour of ECC. This is supported by a study where the application of fiber fraction volume of 2.5% significantly improved the ultimate flexural strength and enhanced the modulus of toughness of ECC mixes [11].

4 Conclusions

This paper presents an overview of the mechanical properties of Engineered Cementitious Composites (ECC). Application of suitable fibers and mineral admixtures enhances long-term mechanical properties and durability of ECC. Mineral admixtures, particularly FA and GGBS, are popular choices as partial replacements of cement in ECC. Enhancement in engineering properties of ECC can be seen even with high percentage replacement of mineral admixture (up to 70%) to cement. In addition to the excellent characteristics of fibers and mineral admixtures in ECC, cost has been one of the major considerations in the selection of these materials. At high temperature, ECC material exhibits a reduction in the compressive strength and tensile strength with decreasing strain-hardening capacity. The mechanical properties of the ECC are highly dependent on combination of the different mineral admixtures, physical properties of the mineral admixtures, and fiber as well as the proportion or volume used in ECC. With the increasing demand for reused and recycled materials, more suitable mineral admixtures can be explored for the development of better performance of ECC, in terms of strength, fire resistance and self-healing capabilities. The mechanical properties of these newly proposed ECCs will certainly need further clarification and the investigation works remain widely open.

Acknowledgements This research was funded by the Universiti Teknologi MARA under grant Research Entity Initiative (REI) 600-IRMI/REI 5/3 (012/2018).

References

1. Leung, C. (1992). Theory of steady state and multiple cracking of random discontinuous fiber reinforced brittle matrix composites. *ASCE Journal of Engineering. Mechanics, 118*, 2246–2264.
2. Şahmaran, M., & Li, V. C. (2009). Durability properties of micro-cracked ECC containing high volumes fly ash. *Cement and Concrete Research, 39*(11), 1033–1043.
3. Li, V. C., Kong, H. J., & Bike, S. G. (2000). *Fiber reinforced high performance concrete material, Ann Arbor*. MI: University of Michigan.
4. Maalej, M., Quek, S. T., & Zhang, J. (2005). Behavior of hybrid-fiber engineered cementitious composites subjected to dynamic tensile loading and projectile impact. *Journal of Materials in Civil Engineering, 17*(2), 143–152.
5. Shi, C., Wu, Y., Riefler, C., & Wang, H. (2005). Characteristics and pozzolanic reactivity of glass powders. *Cement and Concrete Research, 35*(5), 987–993.
6. Ong, K. C. G., Basheerkhan, M., & Paramasivam, P. (1999). Resistance of fibre concrete slabs to low velocity projectile impact. *Cement and Concrete Composites, 21*(5–6), 391–401.
7. Siad, H., Lachemi, M., Sahmaran, M., & Hossain, K. M. A. (2017). Mechanical, physical, and self-healing behaviors of engineered cementitious composites with glass powder. *Journal of Materials in Civil Engineering, 29*(6), 04017016.
8. Şahmaran, M., Özbay, E., Yücel, H. E., Lachemi, M., & Li, V. C. (2011). Effect of fly ash and PVA fiber on microstructural damage and residual properties of engineered cementitious composites exposed to high temperatures. *Journal of Materials in Civil Engineering, 23*(12), 1735–1745.
9. Wang, S., & Li, V. C. (2007). Engineered cementitious composites with high-volume fly ash. *ACI Materials Journal, 104*(3), 233.
10. Yang, E. H., Yang, Y., & Li, V. C. (2007). Use of high volumes of fly ash to improve ECC mechanical properties and material greenness. *ACI Materials Journal, 104*(6), 620.
11. Lee S. W., Oh C. L., and Mohd Raizamzamani M. Z., Evaluation of the Design Mix Proportion on Mechanical Properties of Engineered Cementitious Composites, Key Engineering Materials, vol. 775, pp. 589–595, 2018.
12. Lee, S. W., Oh, C. L., & Mohd Raizamzamani, M. Z. (2019). Mechanical properties of engineered cementitious composites using local ingredients. *Journal of Mechanical Engineering, 16*(2), 145–157.
13. Pourfalah, S. (2018). Behaviour of engineered cementitious composites and hybrid engineered cementitious composites at high temperatures. *Construction and Building Materials, 158*, 921–937.
14. Kim, J. K., Kim, J. S., Ha, G. J., & Kim, Y. Y. (2007). Tensile and fiber dispersion performance of ECC (engineered cementitious composites) produced with ground granulated blast furnace slag. *Cement and Concrete Research, 37*(7), 1096–1105.
15. Mechtcherine, V., de Andrade Silva, F., Müller, S., Jun, P., & Toledo, Filho R. D. (2012). Coupled strain rate and temperature effects on the tensile behavior of strain-hardening cement-based composites (SHCC) with PVA fibers. *Cement and Concrete Research, 42*(11), 1417–1427.
16. Li V. C. (1998). ECC—Tailored composites through micromechanical modeling, fiber reinforced concrete: Present and the future, fiber reinforced concrete: present and the future (p. 64). Montreal: Canadian Society of Civil Engineers.

Comparative Study of Solar Heat Transmission Through Single Skin Façade (SSF) and Naturally Ventilated Double Skin Façade Under Malaysian Climate

Olutobi Gbenga Ayegbusi, Abdullah Sani Ahmad, and Yaik Wah Lim

Abstract Despite existing research on buildings with Naturally Ventilated Double Skin Façade (NVDSF) under Malaysian climate, there are yet unanswered questions. To develop correction factors to calculate overall thermal transfer value of buildings with DSF in Malaysia, this study evaluated the impacts of some design parameters such as window-to-wall ratio (WWR) and glass shading coefficient (SC) on solar heat transmitted through buildings with NVDSF in comparison to that of buildings with SSF on the cardinal orientations. The methodology adopted for this study includes a simulation of a simplified five-storey air-conditioned commercial office building model, using DesignBuilder v5.0.1.024. The results from this study corroborate existing study that a well-designed NVDSF would generally reduce overall solar transmission through the façade which would result in lesser building energy load.

Keywords Solar heat transmission · Single Skin Façade (SSF) · Naturally Ventilated Double Skin Façade (NVDSF) · DesignBuilder

1 Introduction

Modernism ushered in an increase in building design and construction of high-rise buildings with glass façade and postmodern architecture style is mostly characterized by the glass façade envelope which has been found to have energy-related issues [1, 2]. Refes. [3–5] among many other studies that have addressed building energy-use concern as it pertains to solar heat transmission. Similarly, related studies have led to different building design strategies such as overhangs, shading devices, perforated screens, articulated facades, glass curtain walls, Venetian blind application and DSF to improve its building energy performance [3, 4, 6–9].

O. G. Ayegbusi (✉)
Linton University College, Mantin, Negeri Sembilan, Malaysia
e-mail: olutobi@ktg.edu.my

A. S. Ahmad · Y. W. Lim
University Technology Malaysia (UTM), Johor, Malaysia

S. S. Emamian et al. (eds.), *ICSDEMS 2019*,
https://doi.org/10.1007/978-981-15-3765-3_27

Poirazis [10] definedDSF as building typology with two skins—a glazed outer layer and either a glazed or mixed inner layer) arranged to ensure air flows in the cavity. Many studies have been conducted to explore the application of DSF system and the impact on building thermal performance [10, 11]. The results from many of these studies have shown that DSF has good potential and advantages if carefully designed [2, 8, 12]. Under Hong Kong climate, [8, 13] explored different configurations of DSF to identify the most efficient DSF configuration while [14] compared the performance of a DSF with a conventional SSF with absorptive glazing in Hong Kong.

Under the Malaysian climate, NVDSF has been studied and there are few applications on building such as the Securities Commission building [6, 15, 16]. However, the studies presented in [6, 15, 17] only go so far as to consider the indoor air temperature of an office and classroom attached to a ventilated DSF respectively.

With the growing popularity of DSF in Malaysian construction industry, there is the need to develop correction factors to calculate the façade's overall thermal transfer value. Therefore, this study compares the performance of both SSF and NVDSF with varying façade parameters such as window-to-wall ratio (WWR), shading coefficient (SC) and DSF cavity depths on heat transmission through the façade.

2 Method

Different methods have been adopted to evaluate the performance of DSF. Mulyadi [4] presented a study in which field measurement was employed to analyze the temperature distribution and airflow rates in the cavity of an NVDSF for validation purposes. To achieve the present research goal the methodology adopted for this study includes developing a base case model referencing [1, 16, 20] simulation models. The model represents an averagely designed building with respect to building's energy consumption data in Malaysian comparison with a series of models incorporated with a ventilated DSF. The indoor air temperature is maintained at 24 °C with a lighting power density of 21 W/m^2 installed in the conditioned area. The construction detail, thermophysical properties of the simulated models are presented in [18, 19]. The simulated SSF model is as shown in Fig. 1. No shade or window blind is applied on any of the models and hence, the shading coefficient for shading device (SC_2) as given in MS 1525:2014 is not considered in this study. Other simulation assumptions made are detailed out in [18].

DesignBuilder was chosen as the simulation tool for this study. The justifications for the chosen simulation program and its validation is fully discussed in [18]. The in-built simulation manager controls the operations of the simulation modules that coordinate the flow of data to and from the HVAC modules, input data and output data as well.

On the overall, the independent variables include the model's window-to-wall ratio, façade orientation, shading coefficient and the cavity depth for NVDSF cases. These help to understand how different WWR or varying SC would impact heat

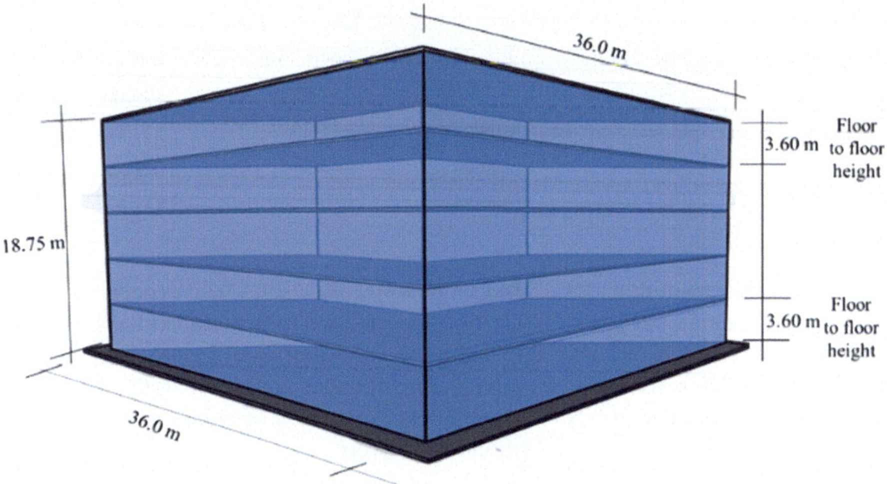

Fig. 1 Base case—non-residential building configurations

transmission. In a comparison of SSF to NVDSF cases, the tested cavity depths (0.2 m, 0.4 m, 0.6 m, 0.8 m) help to understand the impact of the cavity airflow on the hourly solar heat transferred through the ventilated façade system on four different days of the year (March 21, June 22, September 24 and December 21) during the working hours.

To understand how a change in window-to-wall ratio impacts the overall daily heat transfer through façade, varying WWR (30, 50, 70, 90 and 100%) were simulated while the value of the shading coefficients varies from 0.3 to 0.9. The results of the simulation and the analysis are presented in the next section.

3 Result and Discussion

The cumulative solar heat gain recorded on June 22 at the north orientation is a reversal to the amount recorded on March 21, September 24 and December 21 because only the north orientation receives direct solar radiation on this date while other orientations are considered self-shaded. A total of 1141.78 W/m^2 solar heat gain is recorded on the north orientation, the highest of all the orientations. The south, however, transmitted the least solar heat of 780.04 W/m^2 on the same date. Application of a ventilated double-skin façade is not expected to have much impact on the north orientation generally compare to either east or west orientations. Table 1 present the overall daily heat transmission through the cardinal orientation.

The high value of heat transmitted through the glass pane to the conditioned space on December 21 is attributed to the fact that the sun is at the south equinox during this time. Therefore, the combination of cavity ventilation and partition of an NVDSF would prevent direct solar radiation from directly hitting on the conditioned space wall thereby reduces the overall building cooling load. Comparatively, the east,

Table 1 Daily solar heat transmission (W/m²) through the base case model on cardinal orientation

Orientation	Mar-21	Jun-22	Sep-24	Dec-21
East	1210.19	1075.56	1195.20	899.06
North	861.27	1141.78	888.59	627.20
South	924.16	780.04	885.84	1070.31
West	1127.64	1040.97	1085.80	851.86

west and north transmitted 16.0, 20.41 and 41.40% less solar heat compared to the south orientation on December 21. The north orientation mostly transmitted the least solar heat which makes NVDSF less useful on the orientation. In contrast, the east and west orientations transmitted high solar heat with a total of 4370.01 W/m² and 4106.27 W/m², respectively. NVDSF application on these orientations is generally efficient with about 5% to 7% reduction respectively.

3.1 Relationship Between Window-to-Wall Ratio, Shading Coefficients and Heat Gains

Simulation results show a strong relationship between daily heat gain and the WWR with coefficient of determination (R^2) value of 0.99 on the north, south and eastorientations as window-to-wall ratio increases from 30% to 100%. These indicate that the plotted data are predictable with the regression equation. Trend analysis of the simulated data shown in Fig. 2 is a representation of the overall relationship between the

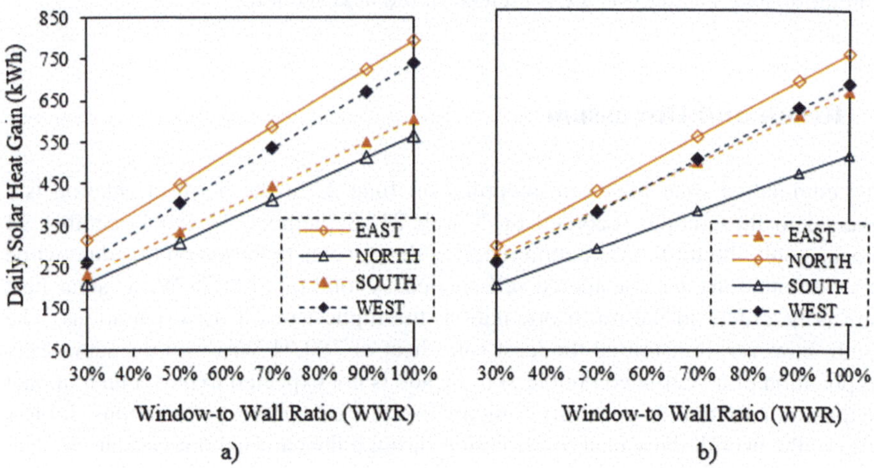

Fig. 2 Impact of window to wall ratio (WWR) on daily heat flow (W/m²) through north orientations of conventional facade

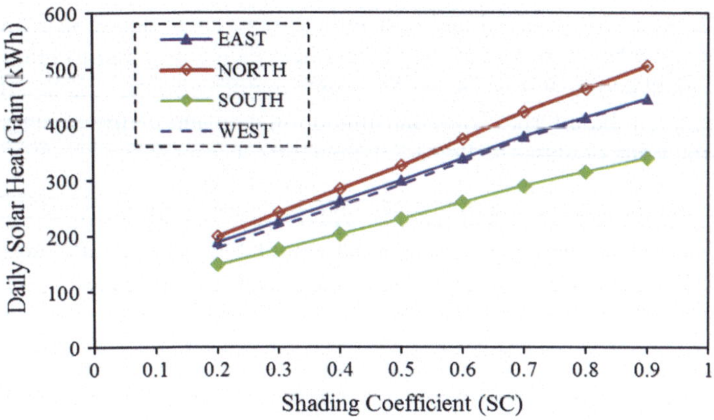

Fig. 3 Impact of shading coefficient (SC) on daily heat flow (kWh) through cardinal orientations of conventional facade

Table 2 Coefficient of determination (R^2) showing the relationship between shading coefficient (SC) and daily heat gain on cardinal orientation

Orientation	East	North	South	West
R^2	0.99	0.99	0.99	0.99

incremental value in the heat gain and the window area throughout the tested days. There is a progressive increment in the daily heat gain as the window-to-wall ratio increases. The progression was predictable with every single increment. The overall results as it relates to total daily heat gain and the resulting cooling load corroborate the existing study on the impact of shading coefficient on building cooling load.

There is an average of 8.89% reduction in the solar transmission between a 90% WWR case and the 100% WWR façade on tested orientations. Hence, 90% WWR is considered preferable to 100% in the Malaysian context as it transmits less overall daily heat gain as well as achieving the necessary transparency.

Similarly, the trend as the shading coefficient varies from 0.2 to 0.9 shows that there is a very strong relationship between the daily heat gain and SC with 0.99 coefficient of determination (R^2) on all the tested orientations (Table 2, Fig. 3). This indicates that more than 99% of the plotted data on cardinal orientations are predictable through the regression equation.

4 Conclusion

The results from this study have shown that the value of heat transmission on any orientation depends on the solar system as an existing study shows. The east orientation transmitted more heat than other orientations on March 21 and September 24.

On June 22, the north received and transmitted more solar radiation while on the 21 December, the South recorded the highest heat gain compared to other orientations. The calculated heat gain as the SC varies confirmed other studies discussed in literature. It can be inferred from the simulation data and analysis that a positive increase in the value of glass shading coefficient did cause an increase in the total heat transmission which is consistent with other earlier related studies. Finally, this study shows that as the buoyancy effect reduces with increase cavity depth, the efficiency reduces. This explains why the facade model with 0.2 m cavity depth outperformed other NVDSF model cases. However, there is less useful additional space with small cavity depth.

References

1. Kannan, K. S. (1991). Thermal characteristics of Malaysia building envelopes. Unpublished Ph.D. thesis, University of Malaya.
2. Li, X., Zhu, Y., & Lin, B. (2006). Experimental Study of Airflow in Naturally Ventilated Double Skin Façade. *The Fourth International Symposium on Computational Wind Engineering., 1,* 267–270.
3. Al-Tamimi, N. A., & Fadzil, S. F. S. (2011). The potential of shading devices for temperature reduction in high-rise residential buildings in the tropics. *Procedia Engineering, 21,* 273–282. https://doi.org/10.1016/j.proeng.2011.11.2015.
4. Mulyadi, R. (2012). Study on naturally ventilated double-skin façade in hot and humid climate. Ph.D. thesis. Nagoya University. Available at: http://ir.nul.nagoyau.ac.jp/jspui/bitstream/2237/18176/1/k9828.pdf.
5. Parra, J., Guardo, A., Egusquiza, E., & Alavedra, P. (2015). Thermal performance of ventilated double skin façades with venetian blinds. *Energies, 8*(6), 4882–4898. https://doi.org/10.3390/en8064882.
6. Bellia, L., Marino, C., Minichiello, F. & Pedace, A. (2014). An overview on solar shading systems for buildings. *Energy Procedia. Elsevier B.V., 62,* 309–317. https://doi.org/10.1016/j.egypro.2014.12.392.
7. Gratia, E., & De Herde, A. (2007). The most efficient position of shading devices in a double-skin facade. *Energy and Buildings, 39*(3), 364–373.
8. Haase, M., & Amato, A. (2009). An investigation of the potential for natural ventilation and building orientation to achieve thermal comfort in warm and humid climates. *Solar Energy, 83*(3), 389–399. https://doi.org/10.1016/j.solener.2008.08.015.
9. Ossen, D. R. (2005). Optimum overhang geometry for high-rise office building energy saving in tropical climate. Ph.D. thesis, University Technology Malaysia.
10. Poirazis, H. (2004). Double-skin facades for office buildings. EBD-R-04/3, division of energy and building design, department of construction and architecture, Lund institute of technology, Lund university, Lund. Available at http://www.ecbcs.org/docs/Annex_43_Task34-Double_Skin_Facades_A_Literature_Review.pdf. [Accessed December 11, 2014].
11. Rahmani, B., Kandar, M. Z., & Rahmani, P. (2012). How double skin facade's air-gap sizes effect on lowering solar heat gain in tropical climate. *World Applied Sciences Journal, 18*(6), 774–778. https://doi.org/10.5829/idosi.wasj.2012.18.06.3184.
12. Elkadi, H. (2006). *Cultures of glass architecture.* Ashgate, Hampshire, UK: Design and the built environment.
13. Chan, A. L. S., Chow, T. T., Fong, K. F., & Lin, Z. (2009). Investigation on energy performance of double skin façade in Hong Kong. *Energy and Buildings, 41*(11), 1135–1142. https://doi.org/10.1016/j.enbuild.2009.05.012.

14. Hashemi, N., Fayaz, R., & Sarshar, M. (2010). Thermal behaviour of a ventilated double skin facade in hot arid climate. *Energy and Buildings, 42*(10), 1823–1832. https://doi.org/10.1016/j.enbuild.2010.05.019.
15. Shafii, F., Arman Ali, Z., & Othman, M. Z. (2006). Achieving sustainable construction in the developing countries of southeast Asia. The 6th Asia-Pacific structural engineering and construction conference, 29–44. https://doi.org/10.1016/j.bmc.2013.06.010.
16. Tang, C. K. & Chin, N. (2013). Passive design, building sector energy efficiency project (BSEEP), Malaysia. Available at: http://www.scpmalaysia.gov.my/images/BSEEP%20Passive%20Design%20Guidebook.pdf.
17. Baharvand, M., Hamdan, M. A., Safikhani, T., & Mirmomtaz, S. M. M. (2013). Thermal performance of tropical atrium. *Environmental and Climate Technologies, 12*(1), 34–40. https://doi.org/10.2478/rtuect-2013-0014.
18. Ayegbusi, O. G. (2019). *Double skin facade overall thermal transfer value correction factor in hot humid climate. Ph.D. thesis.* University Technology Malaysia.
19. Hien, W. N., Liping, W., Chandra, A. N., Pandey, A. R., & Xiaolin, W. (2005). Effects of double glazed facade onenergy consumption, thermal comfort and condensation for a typical office building in Singapore. *Energy and Buildings, 37*(6), 563–572.
20. Chan, A. L. S., & Chow, T. T. (2014). Calculation of overall thermal transfer value (OTTV) for commercial buildings constructed with naturally ventilated double skin façade in subtropical Hong Kong. *Energy and Buildings, 69*(14–21). https://doi.org/10.1016/j.enbuild.2013.09.049.

Responsible Waste Management of Micro, Small, and Medium Enterprises (MSMEs) Toward Jakarta Sustainable Development

David Ronald Tairas, Rosnaini Daga, Nur Fatwa, Lin Yola, and S. Basir

Abstract Jakarta is the capital of Indonesia, the problem of waste cannot be adequately handled by the authorities, and there is still space for improvements to be done. Rapid of globalization and industrial development 4.0. The increase of mass waste production in Jakarta, lacking awareness of the community to dispose of waste in its place coupled with household waste generated by the community and small and medium-sized industrial businesses. This study aims to Responsible Waste Management of Micro, Small, and Medium Enterprises (MSMEs) Toward Jakarta Sustainable Development, in overcoming the waste problem, and it is necessary to have the cooperation of the authority. The community and small and medium industrial entrepreneurs need to be given training and counseling on waste management which can later be reused by small and medium industrial entrepreneurs. The finding of this study is a significant reference to the sustainable development initiative towards a clean and waste-free Jakarta living environment.

Keywords Responsible · Waste Management · Sustainable Development

1 Introduction

Waste is management is still facing essential challenges in big cities in Indonesia. Jakarta is no exception. The rapid increase of population and urbanization have created social and environmental issues [1], which one of them is the increase in waste volume from city dwellers. Currently, the cycle of waste collection ends in the landfill before the selection and disposal process. Jakarta is experiencing the increase of waste issue which is not followed by a sufficient number of landfills. The uncontrolled capacity of the landfill generated due to lacking proper waste management in reducing the waste volume. Therefore, urgent solution is needed. The integration

D. R. Tairas (✉) · N. Fatwa · L. Yola · S. Basir
School of Strategic and Global Studies, Universitas Indonesia, Jakarta, Indonesia
e-mail: davidronaldtairas@yahoo.com

R. Daga
Institut Bisnis dan Keuangan NITRO Makassar, Makassar, Indonesia

S. S. Emamian et al. (eds.), *ICSDEMS 2019*,
https://doi.org/10.1007/978-981-15-3765-3_28

of stakeholders such as the local authorities, private sector, and community to overcome this issue is urged. In this context, the role of community and Micro, Small, and Medium Enterprises (MSMEs) participation is significant. They have opportunities to create innovation in waste management by implementing the 3R (Reduce, Reuse, Recycle), which also will benefit them in the business sector.

The waste management initiative is a key component to implement the sustainability and the clean environment in Jakarta. The early action of waste management needs a tube done by the community and private sectors before sending it to the landfill. The continuous action and enforcement are forms of environmental protection. The waste issue mitigation would be more applicable when the nature of 3R especially on selection, of organic and non-organic, it is well applied to answer the environmental balance is promotion, in fact, the mix of domestic, industrial organic and inorganic waste without any selection process widely occurs in big cities in Jakarta.

In the era of Industry revolution 4.0, an alternative solution using technology and innovation is majorly needed in a waste management system. Table 1 shows the type of waste and the time needed for disposal. The 3R strategic especially the waste selection is a big challenge to all stakeholders. Currently, there are two types of waste recycling; composting (composting fertilizer) and crafts (making various handicraft products made from garbage). In order to increase the awareness of the stakeholders, socialization and education on waste management and its impact on the sustainability and clean environment of Jakarta are a priority. The waste recycling process begins from the selection of organic and non-organic waste to transform into the raw material to be processed into the economic products. The waste recycling aims to use organic waste, by encouraging the agriculture sector and city green program (by using the composting fertilizer). The waste management cycle from the waste shorting until the disposal process and organic fertilizer production could be seen in Fig. 1.

Waste recycling contributes to the positive impact on the economic sector, especially on community income. A study [2] emphasizes on the support on the infestation and the human resources needed in resolving the economic issue. Besides the investment, waste management could be done independently or by implementing the

Table 1 Type of waste and decomposing time

Type of waste	Decomposing time
Paper	2–4 Months
Orange peel	6 Months
Carton box	5 Months
Cigarette filters	10–12 Years
Plastic bags	10–20 Years
Leather shoes	25–40 Years
Nylon clothes	30–40 Years
Styrofoam	Not Decompose

Source Department of Statistics Indonesia, 2019

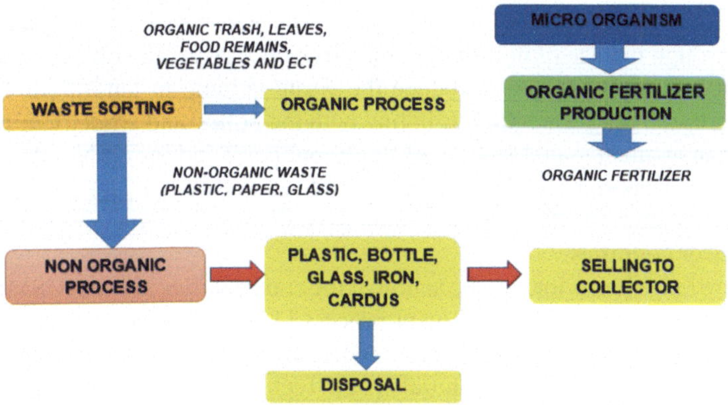

Fig. 1 Waste management cycle. *Source* Department of statistics Indonesia, 2019

waste bank system. The waste bank aims to collect waste after the selection process. Afterward, the waste will be transferred to the temporary waste collection center. The concept of the waste bank represents the banking system in which the system is managed by the staff. The depositor is represented by the community that has been given the bank book. The deposited waste will be weighed and priced in sum number of money before being sold to the collaborated factory.

2 Concept of Waste Management

Environmental protection is basically pinpointed according to the human mindset and behavior. A study [3] reported that the changes in human mindset and behavior on the household management to reduce waste through the community participatory needs to be integrated into the community-oriented waste bank system. Another study [4] on the waste handcraft marketing emphasized that waste need to be recycled. Meanwhile, a similar study [5] on the newspaper recycling stressed out that two of the strategies to overcome the environmental problem are reusing or recycling city dwellers required to transform from the old style of waste disposal, by education and socialization of 3R and the waste bank system [6]. Furthermore, a finding was emphasized that households need to independently perform the action on managing the domestic waste, which in this context is called by community-based waste management [7].

Basically, waste management is formed by the concept of 3R. Technically, the 3R program needs to be done before the disposal process [8]. Beside the 3R, 4R concept is also suggested with the additional action of replace. Meanwhile, the 5R concept was brought with the additional action of replant. However, 3R concept is technically applied in Indonesia. Department of Public Works Indonesia [9], defines the scope of 3R. Reduces aims to mitigate the consumptive lifestyle, and applies the concept of non-disposability to reduce the waste volume. The 'reuse concept' targets to utilize products through the use of non-disposable products. Lastly, the objective to recycle is to do the selection process to reduce the waste volume in landfilled. The 3R is community based program which is in line with the policy and strategy of Indonesia in waste management stated in Law of the Republic of Indonesia Number 18 2008 [10] and act of Department of Public Works Indonesia Number 21/PRT/M/2006 [9]. The implementation of 3R is the key to achieve the waste management framework in big cities or countries.

The knowledge behavior and the action of the communities and the MSMEs actors are essentials in the waste management process [11]. The waste selection (organic and non-organic) aims to be transferred into the waste bank for the recycling process or to transform it into the raw materials [12]. Technically, the waste base products would need marketing strategies. Furthermore, it was pointed out that the marketing process brings the objective of heading the market value of the products, therefore the product could reach the customers [13].

Marketing creates better communication in justifying the products into the customers. The Entrepreneurship Strategy needs to be applied to increasing competitiveness [14]. The identification process of supply and demand is a concept of marketing. The marketing process is mainly based on a supply-demand, market offerings products, services and experiences, customer value and satisfaction, exchanges and relationships, and markets [13]. Other important strategy is marketing management. Marketing management is a form of art and knowledge in market target selection, creation, transportation, and communication of the best value of the product to achieve and sustain the number of customers. The marketing management as a process to improve the efficiency and the affectivity of the marketing activities among the individuals and organizations. Therefore, this study stresses that the implementation of the comprehensive strategy of 3R and the well-managed marketing is a key role in implementing the waste management system in Jakarta.

3 Methods

This study explores the descriptive-qualitative analysis in analyzing the responsible waste management strategy towards the sustainable framework in Jakarta. Jakarta is Indonesia's capital city which covers 661.52 km^2 area, and 10,187,595 populations. The result of the investigation aims to recycle the waste for the income increment for the community, to produce the innovative product with an environmentally friendly recycling concept. Moreover, the study also targets to reduce the urban waste volume toward sustainable development.

This study was conducted in Jakarta which includes 58.91 million units of micro business, 59,260 units of small and 4.987 units of medium, and enterprise. The study analysis the daily waste from 2014–2017 in Jakarta, the approach of waste collection, type of waste (organic, non-organic, and toxic waste), 3R program, and recycle products. The descriptive qualitative analysis discusses the data generated from the Department of Public Works Indonesia, Center for Research and Development of Settlements—Ministry of Public Works and Department of Micro, Small and Medium Enterprises, Department of statistics Indonesia.

4 Results and Research Description

The rapid growth of population, industry, urbanization, and modernization is the root of the waste volume increasing. This situation due to the increasing trend of food demand and other needs, therefore the household waste develops day by day. Table 2 presents the increase in the volume in Jakarta from 2014 to 2017. It could be seen that the rising volume increases consistently. However, waste management is not yet done comprehensively and regularly. The data shows that the collected and transported waste ranges from 91.10% to 95.92%. This situation shows that the role of policymaker MSMEs actors and community is significant to involve in the waste management system to improve the communities' income and sustainable development implementation in Jakarta.

Meanwhile, Table 3 presents the management of Jakarta local authorities to provide the approaches of waste collection from the community and MSMEs. The garbage men are assigned to transport the waste from households and MSMEs to the landfill. They are also responsible for waste sorting when the community or MSMes actors fail to perform it. The knowledge and awareness of the garbage man about the

Table 2 Jakarta waste volume by year

Waste volume	2014	2015	2016	2017
Estimated Waste Production Per Day (m^3)	6.74803	7.04639	7.09908	7.16453
Volume of Waste Trashed Per Day (m^3)	6.21205	6.41914	6.01630	6.87218
Percentage of Transported Waste (%)	9206	9110	8475	9592

Source Department of Statistics Indonesia, 2019

Table 3 Approaches of waste collection in Jakarta by year

Approaches/Tools	2014	2015	2016	2017
The Garbage Man	–	13.708	14.289	10.414
Garbage truck	1.055	1.496	1.749	1.692
Garbage Wheelbarrow	7.422	830	1.092	1.447
Garbage dump	2.554	1.416	1.416	1.099
Heavy equipments	82	1.675	1.990	2.064

Source Department of Statistics Indonesia, 2019

Table 4 Waste groups in Jakarta by year

Waste groups	2015	2016	2017
Organic waste (m^3)	3.45029	3.23377	3.69380
Inorganic waste (m^3)	2.93296	2.74890	3.13996
Hazardous and toxic/B3 (m^3) waste	3588	3363	3842

Source Department of Statistics Indonesia, 2019

type of waste majorly assist in effective implementation of the national policy on the 3R program.

Waste is one of the main potentials to contribute to the environmental positions that include soil, water, and air. In this context, nonorganic waste mostly causes environmental pollution. Table 4 justifies the waste group including organic, nonorganic, and toxic wastes in Jakarta from 2015 to 2017. It indicates that the scale of nonorganic and toxic waste almost reaches the same level as organic waste. This situation is a serious challenge not only to the waste management system in Jakarta but also in the implementation of sustainable development. This study highlights that the increase in waste volume is technically due to the law awareness of community and MSMEs actors to implement 3R program.

If the 3R program is well enforced, the next stage of the waste management cycle is to create and market the recycled products. The examples of the recycled materials are paper and plastic. Human activities in big cities impact, the increase of the paper needs. Paper recycling could be done up to 4–6 times due to the elasticity of the paper fiber. There are ranges of recycled papers that could be categorized into three types: The left-over papers from the factory, the waste papers before use, and the papers after use. Overall, only around 70% of the used papers are recycled, despite the waste papers reaching up to 10% of the total waste. Thus, the recycling strategy of the used papers needs to be maximized. The 3R program, the "reduce" activity could be implemented by avoiding the unnecessary use of papers. In applying the "reuse" activity, the used paper could be reused for other purposes. Lastly, to implement the "recycling, the use of papers could be recycled for artwork. Figure 2 shows the examples of recycled paper products from MSMEs Jakarta.

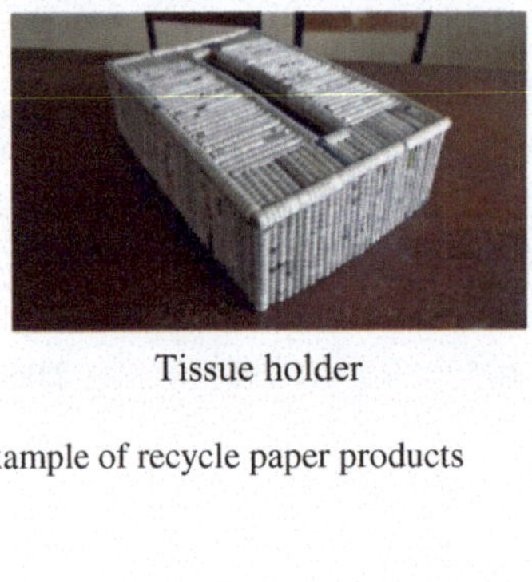

Tissue holder

Shopping Bag

Fig. 2 Example of recycle paper products

Shopping Bag

Lantern Lamp

Fig. 3 Examples of recycle plastic products

Plastic waste management is a challenging process as plastic material mostly decomposed. Plastic has a long carbon decomposition process and unable to be processed by microorganisms. The solution to this issue is reducing the plastic-based products or recycling plastic products into other purposes. In this context, the plastic waste is processed into recycling products such as shopping bags and the lantern lamps (Fig. 3).

The 3R program does not only implement the waste management objective but also is part of sustainable development framework in Jakarta. This is because of the 3R program is the real form of carbon footprint mitigation. Besides, the 3R program improves the community income, encouraging the export of the local products and saving energy from the households to the city scale. Lastly, the 3R program promotes the MSMEs in expanding the business and enterprise on the waste craft products.

5 Conclusions

This study concludes that the comprehensive roles of the local authorities, community, and MSMEs actors are significant in Jakarta waste management. The 3R program needs to be entirely implemented from the early processes of waste selection from the household and MSMEs. The concern of this study is the importance of the awareness among the community and MSMEs actors to apply the waste sorting before the collection and disposal process. Socialization, education, and training on waste management are essentially urged. Moreover, the authority requires presenting the best practice of waste management in order to transform the mindset of the community and the MSMEs actors to be more innovative. The authority also needs to inform them about the consequences of poor waste management and dirty environment. The finding of this study is a significant reference to the community, MSMEs actors, and local authorities to implement better waste management towards sustainable and clean Jakarta.

Acknowledgements This paper is supported by the School of Strategic and Global Studies, Universitas Indonesia. All supports are highly appreciated.

References

1. Yola, L., Siong, H. C. (2018). Impact of urban canyon direction on solar radiation and airflow in hot and humid. *Asian Journal of Behavioural Studies, 3*(3). http://dx.doi.org/10.21834/ajbes.v3i13.146
2. Fatwa, N., Sarita, B., Saenong, et al. (2016). The influence of R&D investment & human capital towards indonesian economic growth rate to address the asean economic community. *Scientific Research Journal* (*SCIRJ*), *IV*(XI). ISSN 2201-2796.
3. Singhirunnusorn, W., Donlakorn, K., & Kaewhanin, W. (2012). Household recycling behaviors and attitudes toward the waste bank project: mahasarakham municipality. *Journal of Asia Behavioral Studies, 2*(6), 35–47.
4. Andini, B. F. (2015). Environmental damage due to plastic waste: cases of citarum river pollution. *Masters Thesis*. Airlangga University Annonim. 2008. Trash Sports. http://www.pusdakota.Jakarta.
5. Siregar, I. (2015). Utilization of newspaper waste for the development of fiber materials. *Journal of Pembangunan Jaya University, 2*, Pembangunan Jaya University.
6. Tallei, T. E., Iskandar, J., Runtuwene, S., et al. (2013). Local community-based initiatives of waste management activities on bunaken island in north sulawesi, indonesia. *Research Journal of Environmental, and Earth Sciences, 5*(12), 737–743.
7. Riswan, S. H. R. (2011). Management of household waste in the South Daha District. *Journal of Environmental Sciences, Environmental Sciences Study Program, Diponegoro University Postgraduate Program, 9*(1), 31–39.
8. Cunningham, D. J. (2004). *Mind, culture, and activity* (No. 2, Vol. 11). Indiana University.
9. Department of Public Works Indonesia, Center for Research and Development of Settlements—Ministry of Public Works. Derived from www.pu.go.id, on 10 September 2019.
10. Law of the Republic of Indonesia No. 18. 2008. Regarding Waste Management Article 1, Article 20 and Article 22.
11. Akhtar, H., & Soetjipto, H. P. (2014). The role of attitude in mediating the effect of knowledge on waste minimization behavior in the terban community, yogyakarta. *Journal of Humans and the Environment, 21*(3), 386–392.
12. Jumar, F. N., Kalalinggi, R. (2014). Household waste management strategies in Lok Bahu village, Sungai Kukung subdistrict, Samarinda City. 2014. *Journal of the Social and Political Sciences Faculty of Mulawarman University, 2*(1), 771–782. Samarinda.
13. Kotler, A. (2001). Marketing principles, Twelfth Edition, Volume 1. Jakarta: Erlangga. 2012. Principles of marketing, Thirteenth Edition, Volume 1. Jakarta: Erlangga., 2014. Principles of Marketing, 15th Edition, Pearson Education Limited.
14. Tairas (2016). The influence of strategic leadership and dynamic capabilities through entrepreneurship strategy and operational strategy in improving the competitive advantage of private universities in Jakarta, Indonesia. *SCIRJ*. *IV*(II), ISSN: 2201-2796.
15. Buchari, A. (2009). *Marketing management and marketing services*. Bandung: Alfabeta.
16. Department of Micro, Small and Medium Enterprises. Derived from http://www.depkop.go.id/.
17. Department of statistics Indonesia. Derived from www.bps.go.id, on 20 September 2019.
18. Lubis, A. (2002). Utilization of waste paper for souvenirs and its marketing. (http://digilib.unimed.ac.id/public/UNIMED-Journal-21355-journal%20dedication%20to%20masociety%20vol.8%20no%2029%20Sept%202002Asri%20Lubis.pdf.

Impact of Urban Configurations on Airflow: Tropical Context Study

Lin Yola, Siong Ho Chin, and Komara Komara Djaja

Abstract The threat of climate change and the Urban Heat Island (UHI) is the environmental challenges to global living development. The UHI phenomenon causes energy demand and social issues in dense cities. The modification of urban surface as the result of the rapid development technically generates the air temperature increase within the urban spaces. The poor urban ventilation further worsens this situation. This study closely looks into the impact of the setting of urban configurations on the airflow within the urban spaces, particularly in the tropical Kuala Lumpur context. ENVI-met V3.1 was used to simulate four urban configurations with two scenarios of canyon directions; East–West and South–North. The results present that the airflow found very weak in four urban configurations. However, the setting of vertical obstruction in the urban configuration helps to improve the ventilation. The study recommends that the canyon feature develops a better airflow in urban configurations. The finding of this study is recommended to be one of the strategies in the Climatically Responsive Urban Configurations framework for urban planners and decision-makers.

1 Introduction

Climate change and urban development are two inseparable topics in the city's sustainability agenda. However, the global agenda on climate change mitigation and adaptation still brings scientific challenges [1]. The big cities play the center role in population growth and urban development that encourage the temperature increase in urban areas. The Urban Heat Island (UHI) is the real form of this issue, as the air temperature in urban area develops higher than the rural areas. A study recorded that the air temperature of Kuala Lumpur rose from 4 °C in 1985 to 5.5 °C

L. Yola (✉) · K. K. Djaja
Department of Urban Studies, School of Strategic and Global Studies, Universitas Indonesia, Jakarta, Indonesia
e-mail: lin.yola@ui.ac.id

H. C. Siong
Department of Urban and Regional Planning, Universiti Teknologi Malaysia, Skudai, Malaysia

S. S. Emamian et al. (eds.), *ICSDEMS 2019*,
https://doi.org/10.1007/978-981-15-3765-3_29

in 2004 [2]. The modification of the urban microclimates is the direct feature that present the increase of the UHI intensity. The UHI phenomenon regularly causes thermal discomfort and urban health to the city dwellers. Consequently, the poor microclimates and the thermal discomfort cause an increase in the cooling load. A study found that every degree of mean ambient temperature increase causes up to 14.2% rise of cooling loads in a typical flat [3]. The issue of climate change and UHI continues to be the challenge in urban sustainability studies.

The previous studies on urban energy budged mostly concerned the climatic features and physical features [4]. The early study [5] justifies the interrelationship and the balance of the modification of the microclimate, thermal comfort, and energy efficiency. Technically, the uncontrolled dense urban development creates many vertical obstructions. The growth of the number and height of high-rise buildings roots the blocked urban wind and natural lighting to reach the ground surface of the urban spaces. In the context of the tropical region, this scenario primarily develops heat storage as one of the sources of the temperature increase. One major component of the microclimates in the urban heat gain system is the urban wind. Urban ventilation plays a significant role in the modification of microclimate and thermal comfort. There are ranges of urban climatic studies that investigated the airflow behavior in urban spaces, mainly urban canyon [6, 7]. Nevertheless, rapid urban development generates the settings of urban spaces and urban form. Hence, this study seeks to explore the further setting of the canyon spaces on airflow behavior.

2 Role of Ventilation in Canyon Spaces of Dense Urban Area

The urban airflow plays a significant role in urban energy budged as heat disperse. Urban airflow also mainly encourages urban thermal comfort [8, 9]. However, a study [10] stressed that with the high level of humidity in the context of the tropical region, the urban airflow significance is still under challenge. The urban surface roughness, such as buildings and topography, mainly determines the airflow behavior that directly contributes to the increase of the UHI intensity [11–14]. The dense city tended to create a lower speed of the airflow on the urban ground surface [9]. A study [13] revealed that the thermal discomfort develops when the airflow is lower than 1.9 m/s in a rough dense urban surface. The airflow velocity increases by the adding height of the surface [7, 11, 14, 15].

Besides the urban roughness, airflow behavior is also mostly influenced by the setting of the urban spaces and built form. Oke T. R. pioneered the study on the relationship between urban canyon and the microclimate [7]. Oke pointed out that more significant Height to Width (H/W) aspect ratio and smaller Sky View Factor (SVF) technically generates the maximum UHI (Eqs. 1 and 2). This study also stresses that H/W aspect ratio determines the airflow regime within the canyon spaces. Another study [11] added to this finding that the airflow develops in the shallow canyon over the deeper canyon. It also expressed the airflow behavior within the different urban canyon spaces, it emphasized that the ground airflow is aligned with the above one when canyon direction faces the urban wind direction, while the spiral vortex develops when airflow above the canyon is angled. Furthermore, the channel effects occur when the canyon direction is parallel to the airflow [11, 16, 17]. The channel effect is emphasized as the primary ventilation system in the dense urban area to reduce heat stress in achieving better urban microclimate and thermal comfort. Therefore, in order to maximize the airflow, it is recommended the canyon parallel to the prevailing wind direction by limiting the deviation to less than 30° [14].

$$dT_{max} = 7.45 + 3.97 * \ln(H/W) \tag{1}$$

$$dT_{max} = 15.27 - 13.88 * SVF \tag{2}$$

However, since the airflow velocity in Kuala Lumpur city center is relatively low, this study sets the canyon direction into East–West and South–North directions to closely seek the impact of the urban canyon on the microclimate. This setting is assigned based on the context of the tropical Kuala Lumpur with the maximum intensity of solar radiation throughout the year [4]. The early study highlighted this situation [18], where the canyon direction towards the urban wind source indicated, generates the impact on the airflow velocity; it, however, also reported that the low velocity is found in Kuala Lumpur ground surface. Besides, this study also underlines the importance of exploring the airflow behavior in other alternatives of urban configurations, by using Oke's urban canyon model as formulated in Eqs. 1 and 2. In this context, the urban configurations with the primary courtyard and canyon geometry were investigated; Courtyard, U, Courtyard Canyon, and Canyon [18]. These urban configurations are also the basic settings found in the modern residential urban blocks [4].

3 ENVI-Met Simulation in Residential Area of Kuala Lumpur

ENVI-met V3.1 Beta, a 3D non-hydrostatic microclimate computer simulation, is the tool used to assess the urban configurations in this study. As validated in an earlier study [19], ENVI-met simulation is a reliable approach to examine the impact of the urban configurations on the airflow. Four urban configurations, Courtyard, U configuration, Courtyard Canyon, and Canyon were simulated to derive the airflow data. In this context, Courtyard Canyon is the existing urban configuration, while the Courtyard, U, and Canyon are the hypothetical urban configurations. The urban simulations set in two sites of the high-rise urban blocks in Kuala Lumpur. The first flat is a 19 story "Bandar Tasik Selatan" with the canyon parallel with the sun path; East-West while the second one is "Surya Magna", a 15 story flat with a canyon direction perpendicular to sun path (North–South). The neighborhood developments include low-rise to high-rise residential buildings. The horizontal surfaces are mostly covered by concrete with 10 m or less of leafless base trees. The investigated focuses study area is the center of the Courtyard, which mostly is used for social activities in the afternoon and evening, and in this context is called receptor.

The four urban configurations are differentiated by the range of SVF value that is generated by the setting of the vertical obstruction. This study, however, remains the H/W aspect ratio constant. The ENVI-met model simulated four urban configurations in two canyon directions; East–West and North–South. The SVF values in Flat Bandar Tasik Selatan are generated by using RayMan analysis from the smallest to highest SVF as follows; Courtyard (0.275), U (0.39), Courtyard Canyon (0.438), and Canyon (0.676). Meanwhile, in Surya Magna the values are as follows; Courtyard (0.611), U (0.694), Courtyard Canyon (0.707), and Canyon (0.793).

The ENVI-met simulation was to simulate the airflow on 21 June 2015 in Kuala Lumpur. The climate and physical data in the configuration editor consist of wind speed (1.4 m/s), wind direction (225 or South–West to North–East), initial temperature (303.15 K), relative humidity (83%), inside temperature of the building (293 K), heat transmission wall (1.94 W/m^2K), roofs (6 W/m^2K), albedo walls (0.3), and roofs (0.5). The size of the simulation sections applied in the simulation of this study is $250 \times 250 \times 30$. In this study, the final extraction of receptor data is the hourly airflow from the Courtyard space. The diurnal data ranges from 7 am to 6 pm, while the nocturnal data is from 7 pm to 6 am. Besides presenting the airflow of the four urban configurations, this study compares the airflow data of two canyon directions; East–West and North–South.

4 Results and Discussion

The ENVI-met simulation is presented as the 24 h airflow data, while the data is also presented in two scenarios of canyon directions; East–West and North–South. In the East–West scenario, the data indicates that the airflow data is deficient (0.01–0.9 m/s). Figure 1 shows that the data is divided into very low and low airflow speeds. It records that the higher rate of airflow occurs in urban configurations with the canyon feature, which in this case includes the Canyon and Courtyard Canyon. The airflow speed indicated to be significant (up to 0.6 m/s). Figure 1 also shows that the Courtyard Canyon and the Courtyard present different trends of diurnal and nocturnal airflows, while the Courtyard and U configuration performed identical airflow trends.

The mean of airflow in four urban configurations stresses this situation (Table 1). The U and Courtyard configurations recorded the deficient speed (0.05 m/s and 0.08 m/s) while the Canyon and Courtyard Canyon perform a better airflow (0.54 m/s and 0.58 m/s). The data also justifies the role of canyon feature in improving the velocity of the airflow in Courtyard spaces of urban configurations. In the scenario of South–North canyon direction, the airflow data is varied among the four urban configurations (Fig. 2).

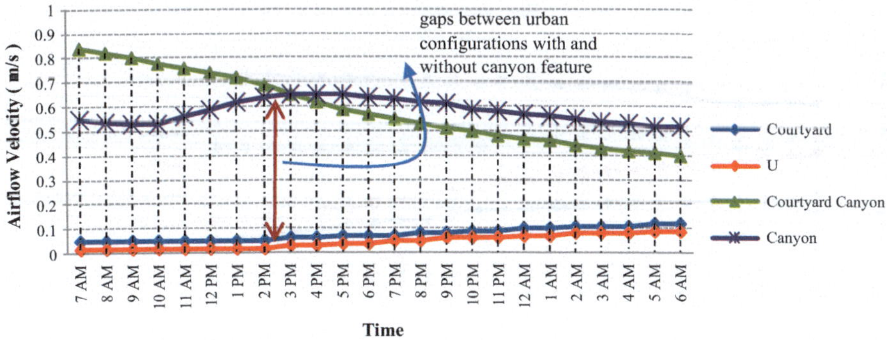

Fig. 1 Hourly airflow in East–West canyon direction

Table 1 Mean of airflow in East–West canyon direction

Urban configuration	Diurnal (m/s)	Nocturnal (m/s)	Mean (m/s)	Remarks
Courtyard (SVF: 0.275)	0.06	0.10	0.08	Low airflow velocity due to the setting of enclosed urban configurations, no help of channel effect occurred
U (SVF: 0.309)	0.03	0.07	0.05	The lowest airflow velocity, the U configuration blocks the source of the wind from South–West
Courtyard Canyon (SVF: 0.438)	0.71	0.46	0.58	The help of the channel effect was well performed due to the small canyon feature in the urban configuration
Canyon (SVF: 0.676)	0.57	0.52	0.54	The lowest airflow velocity found due to the canyon channel effect occurred

In general, the airflow in this scenario recorded as relatively low (0.05 m/s–0.78 m/s). The trend of airflow presented similar characteristics, as the urban configurations with canyon feature indicate a higher velocity (up to 0.78 m/s) compared to the urban configurations without the canyon feature (as low as 0.05 m/s). Apart from the Courtyard configuration, the nocturnal airflow of all the urban configurations reduces after 5 pm. The gaps in the nocturnal airflow are recorded more prominent among the four urban configurations.

Table 2 elaborated on the mean of the airflow in four urban configurations. The

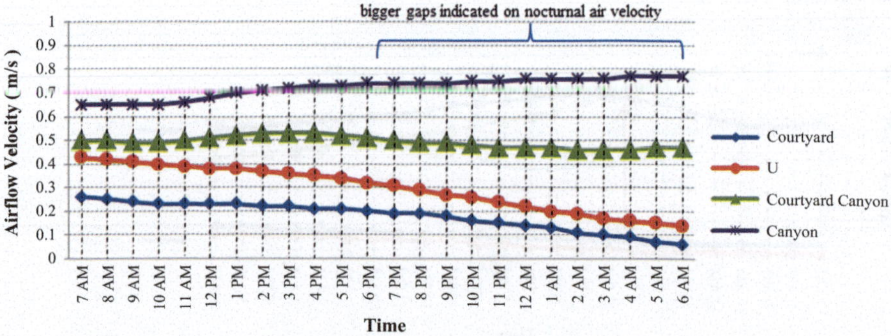

Fig. 2 Hourly airflow in North–South canyon direction

Table 2 Mean of airflow in North–South canyon direction

Urban configuration	Diurnal (m/s)	Nocturnal (m/s)	Mean (m/s)	Remarks
Courtyard (SVF: 0.611)	0.23	0.13	0.18	The lowest airflow velocity as the setting is an enclosed urban configuration
U (SVF: 0.694)	0.38	0.22	0.30	The low airflow, as one of the sides blocked from ventilation access
Courtyard Canyon (SVF: 0.707)	0.51	0.47	0.49	The airflow is better as the channel effect occurs in the small canyon feature
Canyon (SVF: 0.793)	0.69	0.76	0.72	The channel effect found well performed in releasing the heat gain

results show that the channel effect occurs in the urban configurations with canyon feature (Canyon and Courtyard Canyon), while the urban configurations without canyon feature (Courtyard 0.18 m/s and U 0.30 m/s) generated the lower velocity of airflow.

Despite the airflow that was recorded low in all urban configurations, banning the canyon feature proved to create the channel effects that improve the velocity of airflow in the outdoor spaces of the urban configurations.

The simulation data of the two canyon direction scenarios found that the airflow is averagely low in the investigated study. The setting of the dense urban development that plays a role as a vertical obstruction in the sites significantly influences the behavior of the surrounding urban wind. This situation represents the urban wind of Kuala Lumpur with its high-density development in the city center. The results also indicate that there is no significant gap in airflow between the East–West and South–North canyon direction scenarios.

The comparison shows that generally, the nocturnal airflow is recorded lower over the diurnal. In the context of this study, the airflow velocity was physically affected by the setting of the urban configuration. The application of the canyon feature in the urban configuration creates the channel effect that encourages the higher airflow velocity.

5 Conclusion

The increase of UHI intensity in dense urban areas is a challenge to the temperature increase mitigation agenda in tropical cities. The UHI phenomenon causes thermal discomfort and energy demand, which creates the discouragement from the sustainable development agenda. The integration of urban spaces and the built form is one of the significant contributors to modify the heat gain in the urban canopy layer. The primary investigation on the heat gain in the urban canyon spaces highlighted that the maximum intensity of UHI is influenced by the SVF value and the H/W aspect ratio of the canyon.

This study, however, emphasizes that the setting of the urban configurations generates the modification of the microclimate, including the airflow. The results indicate that the existence of canyon feature creates the channel effects in the urban configurations. The canyon feature was highlighted to contribute to the positive impact to encourage the airflow in the canyon spaces. As called Climatically Responsive Urban Configuration, the alternatives suggested in this study are Canyon and Courtyard Canyon. The finding of this research contributes to the reference for planners and decision-makers to strategize the canyon feature to the alternative of urban configuration.

References

1. Masson, V., et al. (2014). Adapting cities to climate change: A systemic modelling approach. *Urban Climate, 10,* 407–429.
2. Elsayed, I. S. M. (2006). The effects of urbanisation on the intensity of urban heat island: A case study on the city of Kuala Lumpur. Ph.D. Thesis. International Islamic University of Malaysia.
3. Liao, F. C., Cheng, M. J. & Hwang, R. L. (2015). Influence of Urban Microclimate on Air-Conditioning Energy Needs and Indoor Thermal Comfort in Houses. Advances in Meteorology, Hindawi Publishing Corporation 2015.
4. Yola, L. (2018). Impact of urban configurations on microclimate and thermal comfort in residential area of Kuala Lumpur. Ph.D. Dissertation. Universiti Teknologi Malaysia.
5. Yola, L. & Siong, H. C. (2017). Impact of Urban Configurations on Air Temperature in Kuala Lumpur, 10th FARU International Research Conference 2017, Vol. 1, pp 418–428. ISBN:

978-955-9027-66-9. Faculty of Architecture, University of Moratuwa, Sri Lanka, 8th–10th December, 2017.

6. Oke, T. R. (1981). Canyon geometry and the nocturnal urban heat island: comparison of scale model and field observations. *Journal of Climatology, 1,* 237–254.
7. Oke, T. R. (1987). *Boundary layer climates.* New York: Routledge.
8. Kang, K., Song, D., & Schiavon, S. (2013). Correlations in thermal comfort and natural wind. *Journal of Thermal Biology, 38*(2013), 419–426.
9. Niu, J., Liu, J., Lee, T., et al. (2015). A new method to assess spatial variations of outdoor thermal comfort: onsite monitoring results and implications for precinct planning. *Building and Environment, 91*(2015), 263–270.
10. Liping, W., & Hien, W. N. (2007). Applying natural ventilation for thermal comfort in residential buildings in singapore. *Architectural Science Review., 50*(3), 224–233.
11. Erell, E., Pearlmutter, D., Wiliamson, T. (2011). Urban microclimate: Designing the spaces between buildings. Earthscan.
12. Gartland, L. M. (2008). Heat Islands: Understanding and mitigating heat in urban areas. Earthscan.
13. Givoni, B. & Noguchi, M. (2004). Outdoor comfort responses of Japanese persons'. Proceedings of the American Solar Energy Society: National Solar Energy Conference 2004. 9–14 July, Portland, OR.
14. Ng, E. (2010). Designing for daylighting—Designing high-density cites for social and environmental sustainability. Earthscan.
15. Givoni, B. (1998). Climate Considerations in Buildings and Urban Design. Van Nostrand Reinhold.
16. Nakamura, Y., & Oke, T. R. (1988). Wind, temperature and stability conditions in an east-west oriented urban canyon. *Atmospheric Environment, 22*(2), 2691–2700.
17. Yuan, C., & Ng, E. (2012). Building porosity for better urban ventilation in high-density cities—a computational parametric study. *Building and Environment, 50*(2012), 176–189.
18. Yola, L. & Siong, H. C. (2018). Impact of urban canyon direction on solar radiation and airflow in hot and humid. *Asian Journal of Behavioural Studies, 3*(13). http://dx.doi.org/10.21834/ajbes.v3i13.146
19. Yola, L., & Siong, H. C. (2017). Computer simulation as an alternative approach in climatically responsive urban configuration study. *Chemical Engineering Transactions, 56*(2017), 505–510. https://doi.org/10.3303/CET1756085. ISSN 2283-9216.

The manufacturer's authorised representative in the EU is Springer Nature Customer Service Centre GmbH, Europaplatz 3, 69115 Heidelberg, Germany. If you have any concerns regarding our products, please contact ProductSafety@springernature.com

Printed and bound by CPI Group (UK) Ltd, Croydon, CR0 4YY
30/09/2026
02225504-0009